Diseases
of
Shade
Trees

Diseases
of
Shade
Trees

TERRY A. TATTAR

Shade Tree Laboratory
Department of Plant Pathology
University of Massachusetts
Amherst, Massachusetts

Revised Edition

ACADEMIC PRESS, INC.
Harcourt Brace Jovanovich, Publishers
San Diego   New York   Berkeley   Boston
London   Sydney   Tokyo   Toronto

56.95

1-7-92

Academic Press, Inc.
San Diego, California 92101

United Kingdom Edition published by
Academic Press Limited
24–28 Oval Road, London NW1 7DX

Library of Congress Cataloging in Publication Data

Tattar, Terry A.
    Diseases of shade trees / Terry A. Tattar. -- Rev. ed.
        p.   cm.
    Includes bibliographies and index.
    ISBN   0-12-684351-1   (alk. paper)
    1. Shade trees--Diseases and pests.   I. Title.
SB761.T37   1989
635.9'77--dc19
                                                        88-32772
                                                        CIP

Printed in the United States of America
89   90   91   92   9   8   7   6   5   4   3   2   1

To
My
Family

Contents

4   Nematodes

5   Viruses

6   Mycoplasmas

7   Seed Plants

8   Leaf Diseases

9   Stem Diseases

10   Root Diseases

11   Rust Diseases

17   Soil Stress

18   Animal Injury

19   Construction Injury and Soil Compaction

20   Chemical Injury

26 *Living Hazard Trees*

Preface

This book was prepared primarily as an introductory work on tree diseases for students and others concerned with the care of shade and ornamental trees. Since my major objective was to develop a text for those without any prior training in plant pathology, much of the technical nomenclature about plant pathogenic microorganisms was not included. Scientific names of trees mentioned were also not included in the text; however, a listing of common and scientific names of these trees is given in Appendix I. A statement about the use of pesticides is given in Appendix II, but specific pesticide recommendations are not included. Additional sources of tree disease information are given in Appendix III.

Diseases of woody plants fall into two major categories: infectious diseases and noninfectious diseases. In Part I, Chapters 2 through 13, the infectious pathogens and the diseases they cause are presented. Chapters on bacteria, mycoplasmas, nematodes, seed plants, and viruses contain discussions of both the nature of these pathogens and the diseases caused by them. The nature of the fungi is presented in Chapter 2, but since the fungi are the most widespread group of infectious pathogens of plants, individual chapters are presented on leaf, root, rust, stem, and wilt diseases caused by fungi. Wound diseases, Chapter 13, which are associated with both fungi and bacteria that invade wounds, are also included in this section. In Part II, Chapters 14 through 23, noninfectious agents and the diseases they cause are presented. Noninfectious agents are separated into environmental stress, Chapters 15 through 17, animal injury, Chapter 18, and people-pressure diseases, Chapters 19 through 22. Diebacks and declines—complex diseases, Chapter 23, which are often caused by a combination of both infectious and noninfectious agents are also included in this section.

Special Topics, Part III, Chapters 24 through 26, includes discussions of nonpathogenic conditions often mistaken for diseases; diagnosis of tree diseases; and living hazard trees, respectively. Although Part III does not present additional shade tree diseases, this information may be of value for those concerned with field applications of the principles given in Parts I and II.

Terry A. Tattar

Preface to Revised Edition

The revised edition of *Diseases of Shade Trees* is organized in a similar manner to the first edition, but information has been added about "new" diseases that have become important since the first edition was published in 1978. In addition, changes in arboricultural practices, such as pruning, tree injection, and fertilization, as well as new products for the arboricultural industry, such as wetting agents and superabsorbant gels, are also presented.

Terry A. Tattar

Acknowledgments

I wish to thank Drs. George N. Agrios, Stanley J. Kostka, and Philip M. Wargo for their valuable assistance with the manuscript and illustrations, and for their general support during the preparation of the revised edition. I also wish to thank my father, Mr. John M. Tattar, for providing drawings for Chapter 1.

Finally, I wish to thank my family for their understanding and encouragement through its completion.

PART I

INFECTIOUS DISEASES

1

Introduction

Of all living creatures, trees are best adapted for life on earth. They live longer, grow taller, and weigh more than all other forms of life. However, they still suffer stress from adverse environmental conditions and from other living organisms. In the natural forest, these pressures on trees serve a positive purpose. They select the most vigorous and healthy individuals, which are the only ones that live long enough to reproduce. As trees become older and lose vigor they become more susceptible to stresses. Eventually they die and rot and so are recycled back into the forest. Thus, the forest "ecosystem" is in perfect balance and stresses on trees have a beneficial role in the continuum of life.

Prehistoric people discovered the multiple uses of trees for fuel, food, shade, beauty, and shelter. People began to see trees as a natural resource, indeed so essential, that sometimes they were worshipped. It was also noticed that some trees were more valuable than others. Selection of trees for man's use had begun. As civilization advanced to recent times man's need for trees appears to have moved along two distinct lines: (1) trees for use as wood products; and (2) trees for shade, beauty, and recreation (Figs. 1.1,1.2). The former generally became the responsibility of the forester and the latter the responsibility of the arborist. There are occasions, however, such as in urban forest recreation areas, when both the arborist and the forester must have the same concerns.

The study of tree health in the forest is called "forest pathology." Its aim is to reduce losses of wood. Shade tree pathology is a study of the health of trees that are grown primarily for shade, ornament, or recreation. Shade trees, in contrast to forest trees, are not usually considered to have potential value for wood products but add value to the areas in which they are growing primarily because of the aesthetic benefits. Shade tree "harvest," therefore, occurs throughout the lifetime of the tree and ceases at its death. In addition, shade trees, due to their longevity, often have personal attachment and historical or even religious significance.

Fig. 1.1 Shade or amenity trees around a home. (Drawing courtesy of John M. Tattar.)

FOREST TREES VERSUS SHADE TREES

In addition to the obvious differences between the management objectives for forest and shade trees there are many other differences between trees growing in the forest and those trees growing around homes, along streets, and in recreation areas. Since the forest is the normal ecological niche for trees it would be expected, in most cases, that more stress would be found in shade trees than on forest trees. The backyard, roadside, city park, and recreation areas are not usually ideal locations for the growth of trees. This increased stress often predisposes shade trees to vigor-related diseases. Thus, a major responsibility of the arborist is to provide the necessary care to maintain vigor and to prevent or alleviate as many of the adverse stress factors as possible from harming the trees. The study of the diseases of shade trees has become a distinct branch of tree pathology because shade trees require specialized maintenance rarely used in the forest, and because shade trees are usually treated as individuals and not as a member of a forest stand.

TREE HEALTH

Understanding the woody plant system is the first step in learning about health of trees. As in any study of disease, the student must begin with a thorough understanding of the "normal" subject. Knowledge of how a tree is put together and functions is essential before one can determine if some part of the tree's system is abnormal or diseased. Often those with little working

Fig. 1.2 Examples of how shade trees contribute to the beauty of homes. (Drawings courtesy of John M. Tattar.)

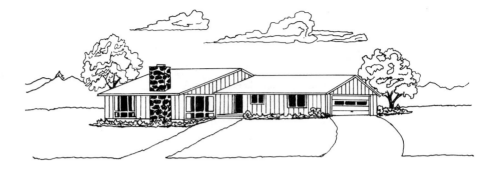

Fig. 1.2 (*continued*)

knowledge of trees attempt to diagnose tree health problems. Many times this results in misdiagnosis due to their ignorance of how a normal tree functions. It is best to have a good background in tree biology and taxonomy before attempting diagnosis and treatment of tree health problems.

The information presented in this text will be most useful for those with a good working knowledge of trees. Once a person is familiar with the healthy appearance of most common shade trees it will not be difficult to detect a disease on those trees in its early stages. The major objective of those working with trees should be to preserve health and to avoid disease. Once a disease has occurred, a tree will be progessively more difficult to treat and to restore to health as the diseased condition becomes more advanced. Therefore, early detection and

Fig. 1.2 (*continued*)

treatment of tree health problems are also key objectives to those working to preserve tree health.

TREE HEALTH CARE

There are several fields related to arboriculture, such as botany, entomology, forestry, microbiology, plant physiology, and soil science, as well as plant

Fig. 1.2 (*continued*)

pathology, that are involved with all the various types of health care problems of shade trees. Since the arborist often needs advice from several different academic specialties there has to be some crossing of fields to provide the needed information. Those who care for trees need to be able to recognize and control tree disorders, regardless of origin, but often do not need to have memorized all the detailed information required to be specialists in a particular field.

CATEGORIES OF SHADE TREE DISEASES

Most shade tree diseases can be placed into two broad categories: (1) infectious diseases, and (2) noninfectious diseases. In addition, there are arthropod-caused injuries and nonpathogenic conditions that are often confused with diseases. However, there often is considerable interaction between all these categories, much to the detriment of the tree and to the confusion of the public. Distinguishing these is the task of the plant diagnostician. In many cases there is also considerable overlap between these categories, but they are still a useful means for understanding the general types of tree health problems.

Infectious Diseases

Disease of plants caused by living pathogens (organisms able to infect plants and cause a plant response) are termed *infectious diseases*. This group of diseases includes most of the well-known disease problems of trees and it has been the group of diseases most studied by plant pathologists. Most of the infectious diseases are caused by microorganisms: fungi, bacteria, nematodes, viruses, and mycoplasmas. But certain pathogenic higher plants also commonly cause tree diseases.

Most of the infectious diseases will be discussed according to the type of pathogen that causes the disease. However, infectious diseases caused by fungi are so numerous that some will be discussed according to the portion of the tree (leaves, stems, or roots) and some will be discussed according to the nature of the pathogen (rusts and wilts). The fungi will first be examined in detail because of their importance as plant pathogens.

Noninfectious Diseases

Diseases of plants caused by nonliving pathogens (nonliving stress factors able to cause injury and death to plants) are termed *noninfectious diseases*. This group of diseases, in general, is not as well known as infectious diseases and often causes considerable problems in disease diagnoses, especially because the tree may have both infectious and noninfectious diseases at the same time. Noninfectious diseases are caused by environmental or meteorological stresses, such as temperature, moisture, solar radiation, animal injury, and human activities known as people-pressures.

Arthropod-Caused Conditions That Mimic Diseases

Insects and mites belong to the group of animals known as arthropods or "jointed-legged animals." Both insects and mites are well-known pests of trees and are such an important cause of animal injury that their studies are separate disciplines. Although obvious insect problems will not be discussed in this text, numerous insect and mite-caused conditions are worth mentioning because they are often confused with common shade tree diseases. Symptoms of such stress usually are apparent on the host but often the responsible insect or mite is obscure. These conditions will be discussed in the sections about diseases that are commonly confused with them.

Nonpathogenic Conditions of Shade Trees

Although most people observe trees every day and feel generally familiar with them, numerous events occur to normal healthy trees that are often overlooked because they occur only periodically. These seemingly abnormal events in the life of the tree are often mistaken for diseases and are brought to the attention of the arborist and plant diagnostician. They are termed *nonpathogenic conditions* since they and their causes are usually not detrimental to the tree in any way, but are recognizable as unique conditions of the tree. Some of these conditions include normal leaf color changes, leaf drop, the presence of lichens, mosses, and sooty molds, and production of flowers or fruit.

SUGGESTED REFERENCES

Nielson, D. G. (1986). Planning and implementing a tree health care practice. *J. Arbori.* **12,** 265–268.
Powell, C. C. (1985). Tree health from top to bottom. *J. Arbori.* **11,** 129–131.
Shigo, A. L. (1982). Tree health. *J. Arbori.* **8,** 311–316.
Tattar, T. A. (1983). Stress management for trees. *J. Arbori.* **9,** 25–27.

2

The Fungi

•

INTRODUCTION

Fungi are nongreen plants that are abundant almost everywhere on the earth. They range in size from those that can only be seen with a microscope to those such as mushrooms and conks that can be quite large. Most species of fungi are beneficial to man. Fungi decompose dead plants and animals and recycle essential minerals back into the soil for reuse by other growing plants. Fungi are also important industrially in the production of antibiotics, cheeses, wines, and a large number of other products. Often fungi serve us as food; yet a certain few species are deadly poisonous if eaten by people. Also, there are a few fungi that are capable of producing diseases in plants and in animals, including man. The species of fungi that can attack woody plants are the major concern of those who work with shade trees.

Fungi are the most commonly encountered group of plant pathogens. Fungi can attack any part of the plant (roots, stems, or leaves), although each species of fungus may attack many parts or only one part. No plants have been found to be resistant to all the various kinds of pathogenic fungi that attack plants. Shade trees are no exception. Many plant pathogenic fungi, however, usually live on dead organic matter and only attack a suitable living host tree when the environmental conditions favor the fungus.

NATURE OF THE FUNGI

Fungi can often be seen as mushrooms on the ground (Fig. 2.1), as conks on a tree (Fig. 2.2), or as mold on cheese or bread, but these rather complex structures are made of much smaller structural units of the fungi, the hyphae. These thread- or tubelike branching strands of the fungus are visible only with the aid of a microscope. When large numbers of hyphae come together, they can

Fig. 2.1 Mushroom of *Armillaria mellea*. (Photo courtesy of Alex L. Shigo, U.S. Forest Service, Durham, New Hampshire.)

produce all the visible structures we commonly associate with fungi. But the hyphae are still the most important structures in considering how the fungi can produce disease in trees. These are the structures that can, and often do, penetrate the defense barriers of the host (tree or shrub) (Fig. 2.3). Once the hyphae have gained entrance into the host the fungus begins to derive its food from the plant and to grow within it. When a pathogen derives food from a plant it is termed an *infection*, and when the pathogen moves within the plant it is termed an *invasion*.

Most fungi can also produce reproductive structures called spores (Fig. 2.4). These consist only of one or a few cells and are invisible to the naked eye (Fig. 2.5). Spores enable the fungus to become dispersed by wind and/or rain to new hosts. The production rate of spores by most fungi is astronomical, often millions per day from a large fruiting body, such as a conk or a mushroom. Spores of plant pathogenic fungi can easily be found in the air around most places on earth. Spores, in general, germinate to produce hyphae, which can also enter susceptible plant species. The size and shape of the spores and the structures that hold them are extremely important in the classification of the fungi. The

Fig. 2.2 Fruiting bodies or conks of (A) *Fomes fomentarius* and (B) *Fomes igniarius* on beech. (Photos courtesy of Alex L. Shigo, U.S. Forest Service, Durham, New Hampshire.)

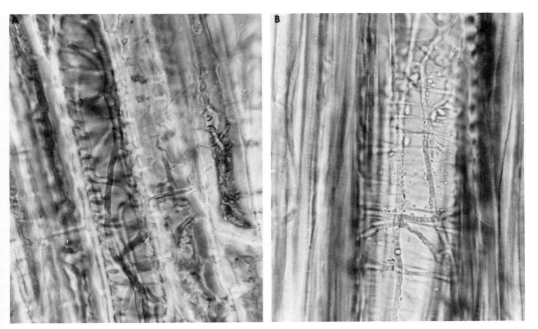

Fig. 2.3A,B Hyphae inside the wood cells of a tree. (Photos courtesy of Alex L. Shigo, U.S. Forest Service, Durham, New Hampshire.)

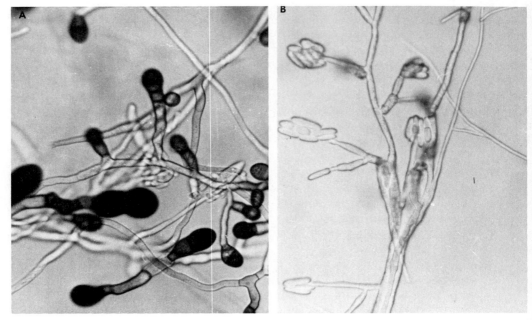

Fig. 2.4A,B   Spores of two fungi greatly magnified (approx. 400×). (Photo courtesy of Alex L. Shigo, U.S. Forest Service, Durham, New Hampshire.)

plant disease diagnostician must often identify the fungus associated with a disease before control measures can be prescribed accurately.

IDENTIFICATION OF FUNGI THAT CAUSE TREE DISEASES

The identification of fungi is a complex process that usually requires microscopic examination and an extensive knowledge of the taxonomy of the fungi. Mycologists devote their whole lives to learning about even limited groups of fungi. Fortunately, the arborist and most other tree specialists can learn to recognize most fungus diseases by appearance of the tree's reaction (the symptoms) and/or in certain cases by the sight of the pathogen itself on the host. The knowledge of these symptoms is the most valuable tool of the plant disease diagnostician in the field. However, with many important fungus diseases, identification of the pathogens is essential, and this can be determined only by microscopic analysis and/or by other microbiological studies in the laboratory. It is the duty of the shade tree specialist, therefore, to know when to submit samples of diseased plant materials to the plant disease diagnostician for

Fig. 2.5 Scanning electron micrographs (SEM) of (A) *Oedocephalon;* (B) *Strumella;* and (C) *Phytophthora cactorum.* (Photo courtesy of Merton F. Brown and H. G. Brotzman, University of Missouri, Columbia.)

identification, what tree tissues are needed in the samples for the particular case, and how to collect and ship the samples.

HOW PATHOGENIC FUNGI ENTER THE TREE

Infectious disease, regardless of pathogen, is a dynamic process involving the interaction of the tree, the pathogen, and the environment. Fungi that can attack woody plants are always at hand, ready and waiting, in the soil, on the bark, and in the air around trees. As the pathogenic fungus attempts to overcome the defenses of the tree, the environmental conditions usually favor either the pathogen or the host. The fungus uses physical and chemical weapons to penetrate the defenses of the host and establish itself within it. Some kinds of fungi enter the plant by growing through natural openings (stomates and lenticels), while other kinds can penetrate directly into the host tissues. But a large number of fungi are able to gain entrance into a plant only through wounds.

Many people do not realize that even very small wounds, such as those from nails driven into a tree, can permit serious diseases to occur in the tree. They also do not realize that stubs of broken branches or improperly pruned branches or mechanical injury from lawn and garden equipment are wounds that are often attacked by fungi. Avoid wounding trees and give immediate, careful attention to fresh wounds! These practices will prevent most infections by fungi on trees and will prevent most infections by other pathogenic microorganisms as well.

HOW FUNGI CAUSE DISEASES IN TREES

Trees respond to the attacks of fungi and other infectious pathogens by placing barriers in front of the invading microorganisms, thus attempting to limit the amount of infected tissue. Unlike an animal, which must kill all invading microorganisms, a plant sometimes can wall off or "compartmentalize" infected tissues. Compartmentalization can take place in any part of the tree: the flowers, leaves, branches, trunk, or roots. The success of the host's physical and chemical barriers against the physical and chemical weapons of the pathogens determines the extent of disease in the tree.

Fungi vary in the location and type of their attacks on the trees. Some attack specific parts, such as leaves, stems, or roots, and others attack certain tissue such as the conducting (vascular) tissue. How the disease looks in the tree will depend, therefore, on the kind of attack by the fungus pathogen. The fungi hyphae can kill cells in the leaves or bark; plug the water-conducting tubes of the wood; decay the wood in the stems, trunk, or roots; cause parts of the tree to grow abnormally or disturb the normal activities of the tree in many other ways. Collective action by many of these microscopic creatures together with the response of the host determine the effects on the tree that we call a disease.

CONDITIONS FAVORING DISEASE BY FUNGI

Fungi need oxygen and organic matter to grow and to cause disease. Given these rather abundant substances, the fungi can produce all the essential materials necessary for their growth, reproduction, and pathogenesis. However, fungi cannot use light to make food as do green plants, and so they grow quite well in heavily shaded or dark places. Fungi are injured by drying because their hyphae are delicate. They grow best where it is dark and moist, with little air movement, thus, where water loss is slight. Fungi can grow over a wide range of temperatures. Some species are adapted specifically for growth in cold climates; others in warm. Most fungi, however, grow best between 68°F (20°C) and 86°F (30°C), which occurs sometime during the growing season of most trees regardless of location.

The conditions favoring growth and reproduction of the fungi are also the conditions that usually favor development of disease in plants. Trees and shrubs growing in heavy shade, or too close together in dense plantings, are more likely to be successfully attacked by fungi than similar plants growing well spaced and in direct sunlight or only partial shade. Shade not only favors fungal growth but heavy shade can also lower tree and shrub vigor by permitting too little sunlight for their green leaves to make food. Trees and shrubs growing in heavy shade or in crowded plantings with little air movement between them also remain wet for long periods after rain. Although the hyphae are very small, they grow and infect plants rapidly when prolonged wet foliage occurs. Sunlight and wind reduce infection by fungi by drying out plants soon after wet periods are over.

SUGGESTED REFERENCE

Brown, M. F., and H. G. Brotzman. (1979). Phytopathogenic fungi: A scanning electron stereoscopic survey. University of Missouri, Columbia. 355p.

3

Bacteria

INTRODUCTION

Bacteria are microscopic, usually single-celled plants that, like fungi, lack chlorophyll, but that are much smaller than fungi. They have a variety of shapes, such as spherical, spiral, or rod-shaped. Some have fine appendages called flagella, which enable a small degree of motility. Bacteria may be found singly but often they appear in loose masses and sometimes in chains. Bacteria multiply extremely rapidly by the simple splitting of one bacterial cell into two cells. This "fission" can occur as often as once every 20 minutes under ideal conditions, so that within a few hours a few bacteria can increase to millions. This way, the bacteria make up for any disadvantage of their small size.

Most bacteria, like most fungi, are beneficial. They are essential in decomposition of organic matter and in soil building, they enable many plants to grow by converting nitrogen available to plant roots, and they are necessary for manufacture of many industrial chemicals. Some species, however, can also produce diseases in man, other animals, or plants. Certain bacteria, like certain fungi, can attack many kinds of woody plants if the conditions for disease are favorable.

HOW BACTERIA CAUSE DISEASE IN TREES

Bacteria do not have any structures to allow them to penetrate directly into plant tissue. Being so small, however, they can enter through natural openings in the plant such as stomates, lenticels, openings at ends of leaf veins (hydathodes), and glands in flowers (nectaries). Wounds are a major site of infection by bacteria, as with fungi. Wounding from pruning, grafting, mechanical injuries, and insects can commonly result in infection of shade trees with bacteria. Bacteria usually are on plant surfaces as contaminants and so they are present when the wound occurs. They also can contaminate the wounding agents such as pruning and grafting tools or insects and thereby can enter every plant that is wounded. Thus, the pathogens from one infected plant can easily be spread to many nearby healthy plants through routine maintenance. Strict precautions, therefore, must be observed to achieve *sterile* tools before grafting and pruning work is begun on each plant.

BACTERIAL DISEASES OF TREES

Fire Blight

Hosts   Fire blight occurs on many species of the rose family including: almond, amelanchier, apple, apricot, blackberry, cherry, chokecherry, cotoneaster, crabapple, exochorda, firethorn (pyracantha), hawthorn, mountain-ash, Kerria, pear, persimmon, plum, quince, raspberry, rose, serviceberry (juneberry), spiraea, and walnut. Resistant species and varieties for firethorn (*Pyracantha cenulata, P. coccinea lalandii,* and *P. fortuneana*) and cotoneaster (*Cotoneaster adapressa, C. dammeri, C. pannosa, C. horizontalis,* and *C. microphylla*) are available and should be requested whenever these hosts are selected for planting. Resistant species and/or varieties may also exist for other hosts as well.

Symptoms   In the late spring, sudden wilting of young leaves, blossoms, and shoots occurs. The affected parts first appear water-soaked and then quickly turn brown then black but remain attached to the tree. These affected areas look as if they had been burned by fire. The infection of shoots may progress down a small branch and start a dark sunken canker in a large branch or in the main trunk (Fig. 3.1). The inner bark in cankered areas also becomes water-soaked first and then turns from green to brown. Milky and sticky ooze that is attractive to insects may also appear on infected bark surfaces especially in warm, wet weather. Wood under cankers is usually brown or black and easily distinguished from adjacent healthy wood. Large sections or the entire tree may become affected if these cankers enlarge and completely girdle branches or the main trunk.

Disease Cycle   Fire blight is caused by the bacterium *Erwinia amylovora* (Fig. 3.2). It remains in a dormant state in blighted twigs and at the edge of cankers during the winter and becomes active during warm spring rains. Bacterial growth results in the production of ooze on twigs and cankers that attract flies, ants, aphids, bees, beetles, and other insects. They carry the pathogen to the blossoms, foliage, and twigs of healthy susceptible trees. The pathogen may gain entrance through wounds caused by insect feeding or through natural openings in the blossoms or leaves. Bacteria are also spread within a mildly infected tree by rain splash of the ooze. Pruning tools can also become contaminated from an infected tree and spread the pathogen to healthy trees through pruning cuts. Infections continue to enlarge until the spring flush of growth stops or about a month after blossoming. The bacteria in the infected tissues generally lie dormant in late summer and remain so until spring.

Fig. 3.1 Fire blight. (A) European mountain ash with fire blight. (B) Close-up of canker margin. (C) Shoot blight symptoms on crabapple. (D) Close-up of branch canker of crabapple. Note papery flaking of outer bark over canker, typical of fire blight.

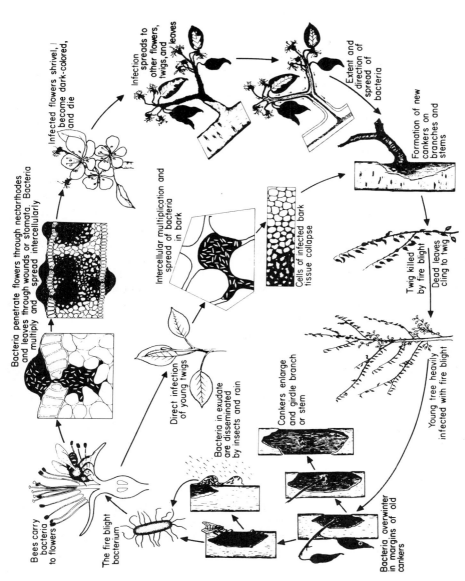

Fig. 3.2 Disease cycle of fire blight of pear and apple caused by *Erwinia amylovora*. [Drawing courtesy of George N. Agrios. (1969). Plant pathology. Academic Press, New York.]

Bacteria penetrate flowers through nectarthodes and leaves through wounds or stomata. Bacteria multiply and spread intercellularly

Infected flowers shrivel, become dark-colored, and die

Infection spreads to other flowers, twigs, and leaves

Extent and direction of spread of bacteria

Formation of new cankers on branches and stems

Intercellular multiplication and spread of bacteria in bark

Cells of infected bark tissue collapse

Direct infection of young twigs

Twig killed by fire blight

Dead leaves cling to twig

Bees carry bacteria to flowers

The fire blight bacterium

Bacteria in exudate are disseminated by insects and rain

Cankers enlarge and girdle branch or stem

Young tree heavily infected with fire blight

Bacteria overwinter in margins of old cankers

Treatment Infected branches should be pruned and cankers on the trunk should be surgically excised during the late summer, fall, and winter. Branches should be cut at least 12 inches (30 cm) beyond the infected area. Surgical excision of a canker is usually performed with a pruning knife although a variety of cutting tools (see Chapter 9, Stem Diseases) have been used to remove infected areas of bark from a tree. The bark is completely removed in small sections over the canker surface, as well as healthy tissue 12 inches above and below the canker and 3 inches on either side. Pruning and surgical tools must be surface-sterilized after each cut. Dipping the cutting tools into any of a variety of common disinfectants, such as a 70% alcohol solution, or a household bleach (sodium hypochlorite) solution (1 part in 4 parts of water) is adequate. Commercial tree paints may be applied to the surface of pruned or excised areas although there is no evidence that this application has any beneficial effect. Occasionally, insufficient tissue is removed from around cankers and excised areas, and pruned faces should be observed periodically during the remainder of the summer to determine if any cankered tissue reappears at the edge of the wound. Any new cankered area should also be excised in the same manner.

Fire blight occurs most commonly in trees with succulent growth of shoots and leaves. Overfertilization, especially with fertilizers high in nitrogen, therefore, should be avoided. Control of insect vectors with regular insecticide sprays during the spring may also be effective in preventing infection. However, if the trees are also used for fruit production care must be taken to avoid killing pollinating insects. Susceptible trees may also be protected with various combinations of the antibiotic streptomycin and various copper-containing fungicides, which are applied several times from early bloom in spring until early summer. Sanitation measures, such as prompt removal and destruction of all infected plant material including recently killed trees and pruned branches from around healthy trees, will decrease the chances of new infections. Fire blight is most famous as a disease of commercial pears and apples and may be more prevalent on shade trees in the vicinity of commercial orchards where these fruit trees are grown.

Crown Gall

Hosts Crown gall is found on a large number of both woody and herbaceous plants and is worldwide in distribution. Many members of the rose family are susceptible but species in other families are also susceptible. Some common woody hosts are almond, apple, apricot, euonymus, fig, grape, peach, pecan, plum, rose, walnut, and willow.

Symptoms Individual or clumps of several small swellings or knobs appear on the surface of the lower stem near the soil line or on roots. These growths usually resemble callus tissue at first but as they enlarge they become more rough, uneven, and brown or black. The galls may also appear higher on the stem and in some cases even have been found in large numbers over most of the

Fig. 3.3 Crown gall. (A) Numerous crown gall infections on weeping willow. (B) Close-up of an individual gall on willow. (C) Close-up of gall on a container-grown rose.

tree (Fig. 3.3). Small woody plants with galls are often stunted but large trees that are infected usually appear normal otherwise.

Disease Cycle Crown gall is caused by the bacterium *Agrobacterium tumefaciens* (Fig. 3.4). This pathogen is often found in soil and usually enters the plant through wounds from pruning, cultivation, or insects. Soon after entering the plant the bacteria stimulate the host to produce large numbers of cells. These cells appear to lose their normal control and continue to divide and expand to abnormal size. The entire area of the stem or root begins to swell as a gall is formed. On small stems a decrease in growth rate results because the gall formation inhibits the flow of water from the roots to the upper parts of the plants. The outer edge of the gall eventually deteriorates due to poor water supply. The pathogen often reenters the soil when the outer edge of the tumor begins to break down; these dead tissues contain the bacteria.

Treatment Crown gall can usually be prevented by avoiding unnecessary wounding, by careful sanitation, and by the use of disease-free stock. Care should be exercised during cultivation to avoid wounding the stem or roots. Pruning and propagation tools should be surface-sterilized frequently. Control of root-chewing insects is also important in areas where the disease is common. Galls should be removed from nurseries by pruning from infected plants whenever possible, or by removal and destruction of the entire plant. Galls on large trees, however, should not be disturbed, since their effect on most trees is minimal. With careful inspection of all nursery stock before planting, there is little chance of planting infected stock. Dipping or soaking root stock in solutions of the antibiotic streptomycin can add protection against infections that occur during transplanting. A bacterium antagonistic to the crown gall pathogen has proven to be a useful biological protectant against crown gall. Root stock or cuttings dipped into a commercial preparation of the bacterium, *Agrobacterium radiobacter*, will not be attacked by the crown gall pathogen.

Infected plants have been soaked or the galls injected with antibiotics or some other chemicals. Such therapy has had only limited success and cannot be recommended.

Galls caused by this bacterium may be confused with those caused by insects. Most insect galls, however, usually do not form near the soil line, and when dissected reveal small holes in the middle where the insect larvae developed. Crown galls, on the other hand, are solid throughout. Burls are sometimes confused with crown gall (Fig. 3.5). Burls are tumors on tree stems that are not associated with any pathogens. The bark of burls usually remains intact in contrast to the deep cracking and breakage of tissue associated with crown gall.

Lilac Shoot Blight

Hosts This disease is found on many woody plants in addition to lilac, such as almond, avocado, apple, cherry, citrus, peach, pear, plum, and rose. Lilac

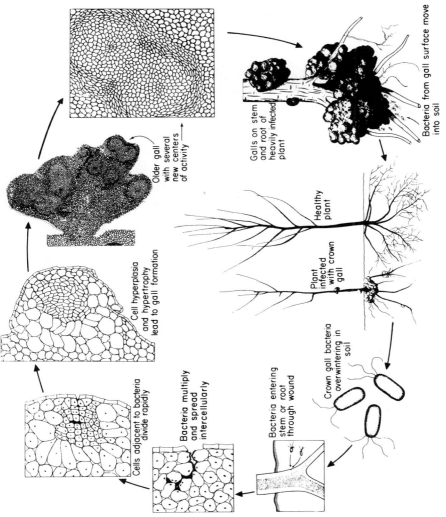

Older gall with several new centers of activity

Cell hyperplasia and hypertrophy lead to gall formation

Cells adjacent to bacteria divide rapidly

Bacteria multiply and spread intercellularly

Bacteria entering stem or root through wound

Crown gall bacteria overwintering in soil

Galls on stem and root of heavily infected plant

Bacteria from gall surface move into soil

Healthy plant

Plant infected with crown gall

Fig. 3.4 Disease cycle of crown gall caused by *Agrobacterium tumefaciens*. [Drawing courtesy of George N. Agrios. (1969). Plant pathology. Academic Press, New York.]

Fig. 3.5 (A) Large stem burl on white ash. (B) Dissection through burl. (Photos courtesy of Alex L. Shigo, U.S. Forest Service, Durham, New Hampshire.)

shoot blight is found more frequently on white-flowered varieties of lilac than varieties with colored flowers but all varieties are susceptible.

Symptoms During the spring, spots appear on young leaves, shoots, and flowers. The spots enlarge rapidly and turn dark brown and then black. Brown streaks can be seen on the wood as the infection spreads to small branches. This disease can be confused with fire blight or late-frost injury. Cankers with accompanying gum exudation are often produced on stone fruit hosts.

Disease Cycle Lilac shoot blight is caused by the bacterium *Pseudomonas syringae* (Fig. 3.6). The pathogen is dormant in infected tissue during the winter. Invasion of healthy tissue from the infected areas, however, can occur whenever the bark of the tree becomes warm, such as by the sun on the southwest side of the tree. Infected areas rapidly increase in size in warm periods just before bud break.

In the spring the pathogen is spread from infected tissues to healthy plants primarily by windblown rain and by insects. It enters the plant through natural openings and wounds, usually on leaves or shoots. The disease is more severe during wet and mild springs than during dry and cold ones. Leaf and shoot infections usually expand, killing branches or causing cankers. Invasion of healthy tissue is slow in summer but faster in warm fall weather when the tree has become dormant.

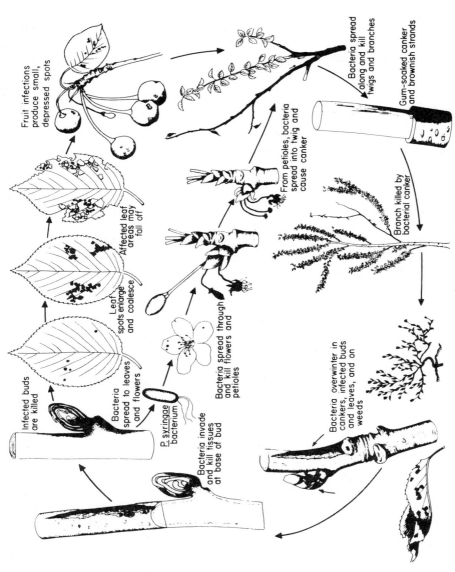

Fruit infections produce small, depressed spots

Affected leaf areas may fall off

Leaf spots enlarge and coalesce

Infected buds are killed

Bacteria spread to leaves and flowers

P. syringae bacterium

Bacteria invade and kill tissues at base of bud

Bacteria spread through and kill flowers and petioles

From petioles, bacteria spread into twig and cause canker

Bacteria spread along and kill twigs and branches

Gum-soaked canker and brownish strands

Branch killed by bacterial canker

Bacteria overwinter in cankers, infected buds and leaves, and on weeds

Fig. 3.6 Disease cycle of bacterial canker caused by *Pseudomonas syringae.* [Drawing courtesy of George N. Agrios. (1969). Plant pathology. Academic Press, New York.]

Treatment   Infected shoots and branches should be pruned out during dry weather. Tools should be surface-sterilized after each cut. After removal of all infected tissues, antibiotic sprays may be applied during spring and early summer to protect the plants from new infections. The use of high-nitrogen fertilizers should be avoided since lush foliage appears to increase susceptibility. Healthy susceptible plants near infected plants should be protected in the same ways.

Bacterial Gall of Oleander

Hosts   Oleander is the only natural host.

Symptoms   Small swellings or knobs appear on the leaves, stems, and flowers. This disease could be confused with crown gall except that roots and any underground portions of the stem are not affected. As the stem growths enlarge they become irregular and blackish. Longitudinal splitting may occur on small branches. Galls on larger branches usually are spherical. Galls on leaves usually occur on the mid-rib and often cause curling or bending. The area around the leaf galls is often yellowed. Plants seldom die from this disease, but infected shrubs are often weak and lose their value as ornamentals.

Disease Cycle   This disease is caused by the bacterium *Pseudomonus sevastanoi* f. sp. *nerii.* The pathogen persists in infected galls and is spread primarily by rain-splash and irrigation water. It can enter susceptible plants through natural openings or through wounds.

Treatment   Diseased parts on older plants and entire young plants should be removed and destroyed as soon as detected. Pruning tools should be surface-sterilized between each cut. Avoid watering by overhead sprinklers, which often serves to spread the pathogen to healthy plants. Antibiotic sprays can be applied in areas where this disease is prevalent.

Walnut Blight

Hosts   Several members of the walnut family including black walnut, butternut, and English walnut.

Symptoms   Brown to black lesions develop on leaves, shoots, and nuts during the spring and early summer. Spots on leaves may join together to cause leaf deformation and tearing of dead tissues, giving the infected leaves a ragged appearance. Infections often extend into twigs with resulting branch dieback and sometimes into larger branches as cankers. Nuts infected early in the season may not remain attached until maturity, but those infected later in the growing season may contain viable and edible nuts.

Disease Cycle   Walnut blight is caused by the bacterium *Xanthomonas juglandis.* The pathogen overwinters in cankers on twigs and branches. During the spring it is carried to healthy susceptible trees by wind-blown rain, pollen, and insects. The bacterium enters the emerging leaves and shoots through stomates, lenticels, or wounds, and often enters nuts at the blossom end from infections in the flowers. The infections often spread through the newly emerged leaf and shoot tissues causing infected parts to have black streaks; later the entire branch may become blackened and shriveled. Twigs and small branches die back if infection continues to spread. However, the infection usually slows as larger branches become involved.

Treatment   Prune out infected branches during dry periods, making sure that cutting tools are surface-sterilized after each cut. Antibiotics may be applied in early spring to protect foliage or nuts where past infections have been known to be severe.

Leaf Scorch

A group of specialized bacteria has recently been shown to cause a vascular disease of some hardwood trees that results in marginal browning or scorch of the foliage on the entire tree (Hearon et al., 1980). These bacteria are much smaller than the bacteria usually associated with plant diseases, such as fire blight and crown gall. They are found only in the xylem and cause dysfunction of the sap conducting cells (Fig. 3.7). Leaf scorch has been found primarily in the mid-Atlantic and southern United States, but mulberry leaf scorch and oak leaf scorch have been found as far north as New York City (Kostka et al., 1986).

Hosts   The following tree species have been affected with leaf scorch: elms (American, Wych, and Siberian), mulberry, red oak, and sycamore.

Symptoms   Marginal browning occurs on leaves, often with a wavy band or halo of yellow tissue on the inside, which borders the normal green tissue near the midvein of the leaf (Fig. 3.8). Initial symptoms may begin in a single branch or in the entire crown. Scorching becomes progressively more severe during the growing season with leaf curling and leaf drop being common by late summer. In cases where the disease has progressed for more than one season twig and branch dieback may occur.

Occasionally leaf scorch is misdiagnosed in the field as moisture stress (see Chapter 16, Moisture Stress) or, in the case of elm leaf scorch, Dutch elm disease (see Chapter 12, Wilt Diseases). Trees under severe moisture stress will sometimes have the brown scorch symptoms but usually do not have the wavy yellow band of leaf tissue inside the brown outer tissue. Trees with bacterial leaf scorch will exhibit scorch symptoms even if well watered throughout the season. Elm trees with leaf scorch will not contain vascular discoloration of the outer xylem as will trees infected with the Dutch elm disease pathogen. Elm trees with chronic leaf scorch, however, are often attacked by elm bark beetles and often die ultimately from the Dutch elm disease.

Control   This disease is most serious on stressed or chronically affected trees where it accelerates a loss of vigor and predisposes trees to attack by secondary

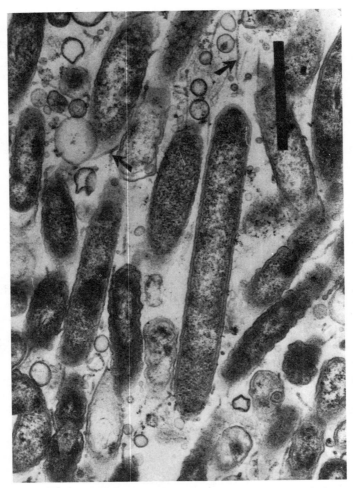

Fig. 3.7 Transmission electron micrograph of leaf scorch bacteria in sap conducting cells of elm. Arrows indicate microfibrils. Bar = 1 m. (Photo courtesy of Stanley J. Kostka, Crop Genetics International, Hanover, Maryland.)

pathogens or insects. Keep trees in a vigorous condition through regular fertilization and watering to avoid drought stress. An insect vector is highly suspect in these diseases although none has been positively identified for leaf scorch of shade trees. Chronically infected trees, therefore, could be a source of the pathogen for nearby healthy trees and should be removed. Preliminary research with the microinjection of oxytetracycline antibiotics has been encouraging, and such therapy may be available for this disease (Kostka *et al.*, 1985).

Fig. 3.8 Leaf scorch symptoms of American elm (A); and marginal necrosis (N) and bounding chlorotic halo (arrows) on mulberry (B). (Photos courtesy of Stanley J. Kostka, Crop Genetics International, Hanover, Maryland.)

LITERATURE CITED

Hearon, S. S., J. L. Sherald, and S. J. Kostka. (1980). Association of xylem-limited bacteria with elm, sycamore, and oak leaf scorch. *Can. J. Bot.* **58**, 1986–1993.

Kostka, S. J., T. A. Tattar, and J. L. Sherald. (1985). Suppression of bacterial leaf scorch symptoms in American elm through oxytetracycline microinjection. *J. Arbori.* **11**, 54–58.

Kostka, S. J., T. A. Tattar, J. M. Sherald, and S. S. Hurtt. (1986). Mulberry leaf scorch, new disease caused by a fastidious xylem-inhabiting bacterium. *Plant Dis.* **70**, 690–693.

SUGGESTED REFERENCES

Anonymous. (1970). Blight of pears, apples, and quinces. *U.S., Dep. Agric., Agric. Res. Serv. Leafl.* **187**, 1–6.

Baker, U. F. (1971). Fire blight of pome fruits: The genesis of the concept that bacteria can be pathogenic to plants. *Hilgardia* **40**, 603–633.

Braun, A. C. (1952). Plant cancer. *Sci. Am.* **186**, 66–72.

Davis, S. H., and J. L. Peterson. (1976). Susceptibility of cotoneasters to fire blight. *J. Arbori.* **2**, 90–91.

Fahy, P. C., and G. J. Persley. (1983). Plant bacterial diseases—A diagnostic guide. Academic Press, New York, 393p.

Hammerschlag, R., J. Sherald, and S. Kostka. (1986). Shade tree leaf scorch. *J. Arbori.* **12,** 38–43.

Lambe, R. C., and G. H. Lacy. (1982). Crown gall. *Am. Nurseryman* **155,** 113–114.

Lippincott, J. A., and B. B. Lippincott. (1975). The genus *Agrobacterium* and plant tumorigenesis. *Annu. Rev. Microbiol.* **29,** 377–405.

Moller, W. J., and M. H. Schroth. (1976). Biological control of crown gall. *Calif. Agric.* **30,** 8–9.

Schaad, N. W. (1980). Laboratory guide for identification of plant pathogenic bacteria. American Phytopathological Society, St. Paul, MN, 72p.

Schroth, M. N., W. J. Moller, S. V. Thompson, and D. C. Hildebrand. (1974). Epidemiology and control of fire blight. *Annu. Rev. Phytopathol.* **12,** 389–412.

Sherald, J. L., S. S. Hearon, S. J. Kosta, and D. L. Morgan. (1983). Sycamore leaf scorch: Culture and pathogenicity of fastidious xylem-limited bacteria from scorch-affected trees. *Plant Dis.* **67,** 849–852.

Van der Zwet, T., and H. L. Keil. (1979). Fire blight—A bacterial disease of rosaceous plants. U. S. Dept. Agric., Agric. Handbook #510, 200p.

4

Nematodes

INTRODUCTION

Nematodes are animals lacking internal skeletons (invertebrates) that are common inhabitants of most bodies of water and are abundant in most soils. Nematodes are also called eel-worms, roundworms, nemas, or simply worms, but they are much more primitive than the earthworms commonly found in gardens or lawns. They are covered with a thin body wall composed of a cuticle, epidermis, and muscles. Inside the body wall the digestive, reproductive, nervous, and excretory systems are contained. The body wall is referred to as a hydrostatic skeleton because it is held rigid by pressure from the interior organs. Their simple anatomy allows nematodes to live without any organized respiratory or circulatory systems. They vary in size from almost 2 feet (approx 60 cm) to just a few hundreths of an inch (approx 1 mm) in length. Most plant-parasitic nematodes are between 0.5 and 5 mm long.

PLANT PARASITIC NEMATODES

Most species of nematodes are free-living particulate feeders that subsist on bacteria, fungus spores, small soil animals, and organic matter. However, some species attack man and animals and other species attack plants. Nematodes are known to cause some of the most severe diseases of man and domestic animals, such as elephantiasis, hookworm, pinworm, and trichinosis, but their importance as pathogens of plants, especially shrubs and trees, is just beginning to be realized. Plant-parasitic nematodes can be separated from other nematodes by the possession of a hollow daggerlike feeding structure known as a stylet (Fig. 4.1). The nematode uses the stylet like a hypodermic needle to pierce the cell walls of plant tissue and either feeds directly on the injured cells or moves into the plant through the newly created wound. Plant-parasitic nematodes usually are near the lower limit of size for nematodes, which could explain why they have often been overlooked as causal agents in tree diseases. In addition, the

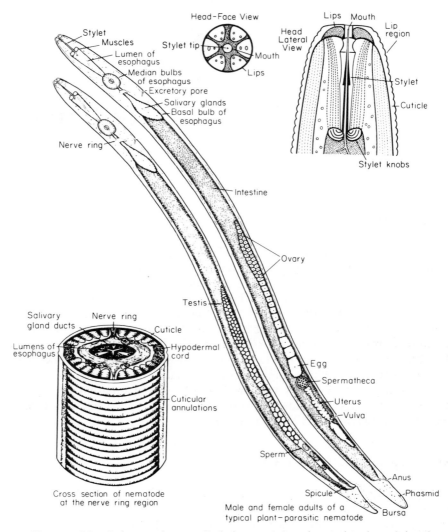

Fig. 4.1 Morphology and anatomical characteristics of typical male and female plant-parasitic nematodes. [Drawing courtesy of George N. Agrios. (1969). Plant pathology. Academic Press, New York.]

aboveground symptoms of nematode attack, such as reduced growth and chlorosis, are typical of a large number of infectious and noninfectious root diseases.

Nematodes can attack almost any part of the plant but the major area of attack with woody plants is the root system. The area of most nematode feeding is the tips of small feeder roots. Nematodes are normally classified by their feeding habits, their movements, and the type of injury they cause. Some species feed only on the surface of the root (ectoparasites) while others move within the root (endoparasites). Nematodes may move along a root or move from root to root (migratory) or may remain in one location (sedentary). Migratory ectoparasites cause stunting of feeder root formation and growth, which results in blunt or "stubby" roots. Migratory endoparasites cause death of large numbers of cells by their feeding and movements, which results in sunken, dead areas or lesions. Sedentary endoparasites cause small swellings or knots to form on the roots.

Nematodes live on the thin film of water that surrounds each soil particle and like the fungi they are very sensitive to desiccation. They can only move through the soil at a very slow rate but can be transported great distances in soil or in surface water.

LIFE CYCLE

Plant-parasitic nematodes have a relatively short life cycle, ranging from about 2 weeks to 2 months, which enables the population to increase rapidly under favorable conditions. The female nematode lays eggs, which shed or mold their cuticles once while still in the egg and then hatch to become motile larvae. As the immature nematodes grow they must molt their cuticle three more times. The juveniles, however, are capable of feeding on plant tissues as soon as they hatch but do not reach sexual maturity until after the fourth molt. Male and female nematodes mate and a new cycle is begun.

Nematodes have also developed some special modes of reproduction. In some types of nematodes both male and female sex organs are found on all individuals. These hermaphrodites are usually not self-fertile and must mate, but each individual lays eggs instead of only half the population as when males and females are separate. Another adaptation in some species is the elimination of males entirely (parthenogenesis), where all members of the species are females and reproduction is asexual.

HOW NEMATODES CAUSE TREE DISEASE

Parasitic nematodes usually attack the most important water and nutrient absorbing area of the root, the small feeder roots. Although the feeding of one nematode may seem insignificant when dealing with a tree, many thousands or millions of these animals attacking the entire root system can be very damaging. The feeding of nematodes can result in dead areas or lesions, gall formation, or lack of feeder root formation on the roots (Fig. 4.2). The net result in all cases is a decreased flow of water and nutrients to the aboveground parts of the plants. The root system also does not fully develop and appears sparse and stunted when compared to healthy plants (Fig. 4.3). The growth of the whole plant is generally slowed, resulting in a runt or dwarf that often has yellow-green foliage.

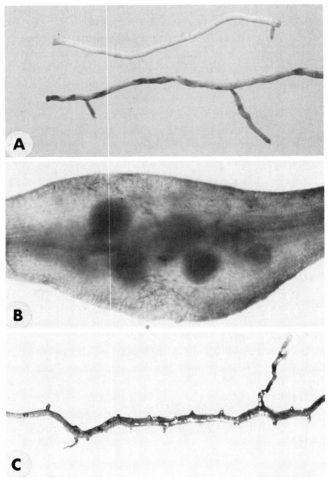

Fig. 4.2 Damage to roots of trees from plant-parasitic nematodes. (A) Comparison of healthy (above) and injury from the root lesion nematode *Pratylenchus penetrans* (below) on yellow poplar. (B) Root knot gall on catalpa. (C) Stubby root on slash pine caused by *Belonalaimus longicaudatus*. [Photo courtesy of John L. Ruehle. (1973). *Annu. Rev. Phytopathol.* **11**, 99–118.]

Nematodes can also serve as predisposing agents for other root pathogens, such as root rot fungi, that can more easily attack roots already wounded by nematodes. They are known, in some cases, to be able to transmit viruses to healthy plants on their stylets. Nematodes are often considered part of a series of infectious and noninfectious stresses (disease complexes) that result in declines of many backyard and roadside trees.

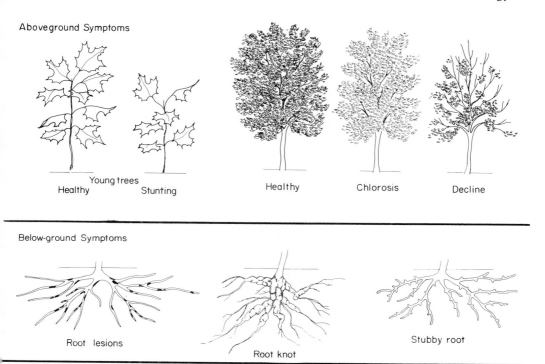

Aboveground Symptoms

Young trees
Healthy   Stunting   Healthy   Chlorosis   Decline

Below-ground Symptoms

Root lesions   Root knot   Stubby root

Fig. 4.3 Effects of nematodes on trees.

NEMATODE DISEASES OF TREES

Many species of nematodes attack the roots of trees and shrubs but plant-parasitic nematodes can be placed into three broad categories by their mode of attack: (1) root knot, (2) root lesion, and (3) stubby root nematodes. Root knot nematodes (sedentary endoparasites) do not kill tissues on the roots but cause swellings or galls to form on the root. Root lesion nematodes (migratory endoparasites) kill many cells as they feed inside the root tissues and cause the formation of dead areas or lesions on the root. Stubby root nematodes (migratory ectoparasites) feed outside on the tips of feeder roots and cause them to stop growing.

Root Knot

Hosts All woody and herbaceous plants are susceptible; some of the common woody plant hosts include althea, boxwood, buddleia, cape jasmine, catalpa, clematis, deutzia, euonymus, fig, dogwood, forsythia, maple, nandina, peach, pine, privet, rose, spirea, and weigelia. Although many woody plants can have root knot there are several types of nematodes that can cause this disease. In some cases a species of root knot nematode may attack only one or a few species of woody plants while in others a large number of hosts may be attacked.

Symptoms   The aboveground symptoms of root knot are usually similar to any one of many general types of root stress. The root knot infected plant often appears as though it is not getting sufficient water although soil moisture is not limiting and/or it may appear to be under fertilized or lacking some other essential mineral, although a soil analysis indicates a normal nutrient balance. In many cases, however, the tissue of root knot affected plants are potassium deficient. Many people may simply observe the foliar symptoms and forego any further investigations, concluding that the problem is caused by drought or some type of mineral disturbance. The key to diagnosing nematode infection or any other apparent root stress problems, however, lies in a close examination of the root system. The roots of root knot infected plants are swollen with small galls (Fig. 4.4). The amount of the gall is directly related to the severity of the disease.

Disease Cycle   Root knot nematodes may overwinter in the soil as eggs or juveniles. The juveniles can infect roots, and the eggs will hatch whenever soil becomes warm regardless of season. Both male and female larvae enter a small root of a susceptible plant and begin to feed on the vascular system. Root knot nematodes do not kill the tissue but cause the developing vascular tissue around the feeding area to enlarge and produce large, unvacuolated "giant cells." The formation of giant cells is solely to benefit the nematode and provide the necessary nutrition for the nematode growth and subsequent production of large numbers of eggs. The root begins to swell in this area and continues to enlarge as the nematodes complete their larval stages and become adults. Mating occurs shortly after they reach the adult stage and the males then leave the root. After fertilization the female's body becomes progressively enlarged with developing eggs, causing the further enlargement of the root. Large amounts of the water and nutrients absorbed by the feeder roots and moving in the vascular system toward the aboveground parts of the plants are being channeled into the root knot nematodes throughout their development. When the eggs are mature the female causes a rupture in the side of the root next to her body and releases the eggs outside the root into the soil to begin a new disease cycle.

A few root knot nematodes will have little effect on a healthy plant, especially a woody plant with a large root system, but large numbers of galled roots can affect the health of a plant regardless of size. However, owing to their developing root systems young plants are more susceptible to serious injury by root knot nematodes than older plants.

The mobility of root knot and other nematodes is small. On their own they can infect only plants very close to the location of their birth. They can, however, be moved great distances in soil and in infected plants, and can also be transported

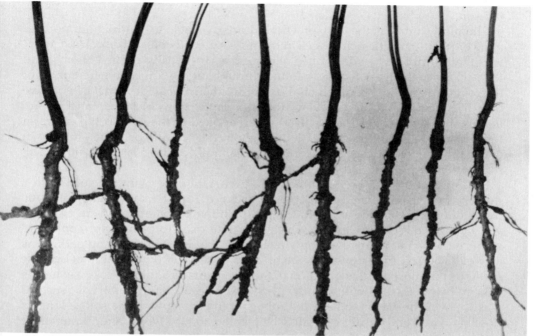

Fig. 4.4 Root knot on catalpa caused by the root nematode *Meloidogyne incognita*. (Photo courtesy of USDA Forest Service.)

locally in surface water especially used for irrigation. Once established in a field, nursery, or around a shade tree they are usually able to persist and often to increase their population size. In the absence of a susceptible woody plant they may remain established on weeds or crop plants for an indefinite period.

Treatment   Root knot is most troublesome on young trees and shrubs. Many of these plants are already infected with root knot when purchased by the homeowner, arborist, or municipal tree official. When obtaining root stock for nursery use or for outplanting, care should be taken to ensure that all plants are free of root knot and other plant parasitic nematodes. These infected trees will make poor shade trees that will probably not survive to maturity. Purchase plants only from reputable dealers who will guarantee their products. If in doubt, contact your county agent or the extension plant pathologist at your state university for advice. An added danger of purchasing nematode infected plants is the possibility of establishing these pathogens in your nursery or backyard.

High populations of root knot and other plant-parasitic nematodes already exist in many areas. Care to avoid infected plants will not be sufficient to control the problem in these areas. Many valuable shade trees and shrubs are already infected, and some type of therapy is needed to help them regain normal vigor. In these cases, efforts must be taken to decrease the population of nematodes in the soil.

Chemical control is the only practical means available to control high populations of plant-parasitic nematodes in the nursery, roadside, or backyard. Chemicals that control nematodes, called nematicides, are usually applied directly into the soil. Since the nematodes are found throughout the soil zone where most of the roots are growing (top 18 inches or 45 cm), the nematicides must be effective over a large volume of soil even around a single tree. Most nematicides are liquids that become gases (volatilize) when applied to the soil. In the gas form the nematicide can diffuse quickly throughout the soil and be quite effective in eradicating nematodes. This form of chemical control is known as soil fumigation.

Soil fumigation may be performed as a preparation for planting, such as in a nursery bed, along the roadside, or a backyard, or it may be a therapeutic treatment for an already infected tree. In planting preparation where only soil will be treated, soil fumigants, which are quite toxic to plants, are applied to sterilize the soil to remove not only nematodes but most other living pathogens as well. The soil fumigants may be applied in the soil by a variety of specific methods but most involve injection of the fumigants as liquids at least six inches (15 cm) below the soil with some type of seal placed on top of the soil (such as plastic sheets or water) to hold the fumigant in contact with the soil as long as possible. Trees and shrubs can safely be planted about 2 weeks after fumigation. Soil fumigants used for therapy, on the other hand, are not toxic to plants when used according to the instructions on the label. These fumigants can be applied as a drench over the roots within the drip line of a tree or shrub or in a granular formulation. This type of soil fumigant may also be used to protect new trees or shrubs by applications at the time of planting.

Root Lesion

Hosts A large number of woody plants are commonly attacked by root lesion nematodes and most woody plants are susceptible to attack by at least one of the several species of nematodes that cause root lesions.

Symptoms The aboveground symptoms are often quite similar to those caused by root knot, an appearance of mild drought, or some type of mineral deficiency. As in detection of root knot, examination of the root system is the key to detecting the injury due to root lesion nematodes. The root system of injured plants is poorly developed with a noticeable lack of most of the small feeder roots. Closer examination will often reveal dead areas (lesions) on many of the smaller roots. The greater the amount of the root area with lesions the greater the amount of injury. Diagnosis of injury from root lesion nematodes however, is often difficult without the aid of a microscope, and it is best to seek the advice of a plant pathologist if you suspect this problem. The plant

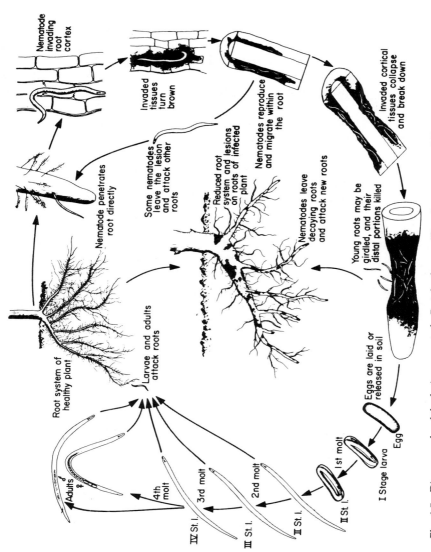

Nematode invading root cortex

Invaded tissues turn brown

Nematode penetrates root directly

Some nematodes leave the lesion and attack other roots

Nematodes reproduce and migrate within the root

Invaded cortical tissues collapse and break down

Root system of healthy plant

Reduced root system and lesions on roots of infected plant

Nematodes leave decaying roots and attack new roots

Young roots may be girdled, and their distal portions killed

Larvae and adults attack roots

Eggs are laid or released in soil

Egg

I Stage larva

1st molt

II St. l.

2nd molt

III St. l.

3rd molt

IV St. l.

4th molt

Adults ♂ ♀

Fig. 4.5 Disease cycle of the lesion nematode *Pratylenchus* sp. [Drawing courtesy of George N. Agrios. (1969). Plant pathology. Academic Press, New York.]

pathologist will closely examine the root system for injury and can also determine the type and number of plant parasitic nematodes in the soil around the roots.

Disease Cycle Lesion nematodes overwinter as eggs, larvae, and adults in soil or infected roots (Fig. 4.5). Larvae and adults enter roots and move through the cortical tissues along the small feeder roots. Lesion nematodes, therefore, are endoparasitic and their feeding habits result in large numbers of dead cells on the small roots of the plant. The general effect of lesion nematodes on the plant health is similar to the effects of root knot although the mode of attack of lesion nematodes is quite different. As more and more of the small absorbing roots are injured by nematode feeding injury the plant slows in growth and declines in vigor. The population of nematodes will continue to increase as long as sufficient susceptible plants are available and favorable environmental conditions persist. The plants are often attacked by other root pathogens, usually soil fungi, that are able to infect the roots through the wounds made by the nematodes or are most easily able to attack a plant in a weakened condition. These secondary pathogens are often diagnosed as the cause of death of the shrub or tree while the primary pathogen, the nematodes, are overlooked.

Treatment Injury from root lesion nematodes can be prevented and controlled in a similar manner as that from root knot. Attack by root lesion nematodes is also most serious on young trees and shrubs. Similar precautions should be undertaken to obtain plants free of plant-parasitic nematodes. Soil fumigation is also recommended for controlling high populations of root lesion nematodes in the soil both for preparation of areas for tree planting or as a therapeutic measure around an affected shade tree.

Stubby (Corky) Root

Hosts Most woody plants can be attacked by stubby or corky root nematodes.

Symptoms The aboveground symptoms of stubby root attack, such as stunting and chlorosis, are similar to those of other nematodes and also of numerous general root disorders. The root system is sparse with very few feeder roots (Fig. 4.6). No necrosis is evident although the roots may appear slightly darkened. Small rootlets are blunt and corky with small bumps on the surface.

Disease Cycle Stubby root nematodes are ectoparasitic and all stages overwinter in the soil. In the spring the nematodes begin to feed near the root tip (Fig. 4.7). Their feeding causes the root tip to stop growing, and it becomes thickened or "corky." Secondary rootlets start to emerge but are also attacked and only form small bumps on the root. The growth of the plant slows as the number of affected roots increases.

Treatment The same control measures described for root knot and root lesion nematodes, such as obtaining nematode-free stock, sanitation, and soil fumigation are also recommended for stubby root nematodes.

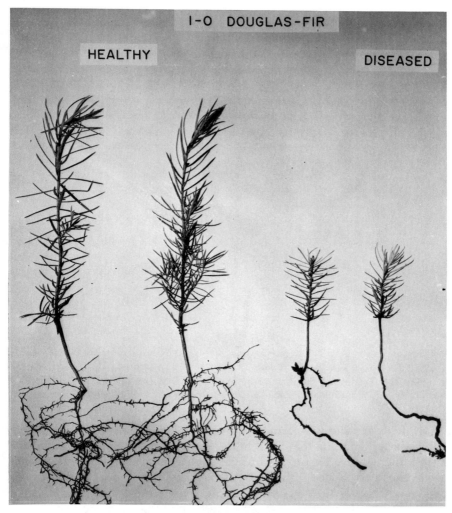

I-0 DOUGLAS-FIR

HEALTHY

DISEASED

Fig. 4.6 Comparison of healthy Douglas-fir seedlings and those with stubby roots caused by feeding by *Xiphinema baleri*. (Photo courtesy of Jack R. Sutherland, Canadian Forestry Service, Victoria, British Columbia.)

Pine Wilt and the Pine Wood Nematode

It has recently been shown that nematodes can cause a vascular disease of conifers that has reached epidemic proportions in Japan (Wingfield *et al.*, 1982). In Japan millions of native Japanese pine trees have died in forests and as shade trees as well. The disease was typified by a sudden collapse or wilt of the entire

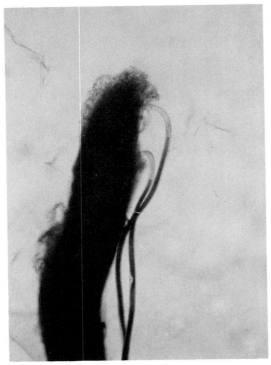

Fig. 4.7 Feeding at the root tip by the stubby root nematode X. *bakeri*. (Photo courtesy Jack R. Sutherland, Canadian Forestry Service, Victoria, British Columbia.)

tree in a few weeks and was termed *pine wilt*. The pathogen that caused pine wilt in Japan was a nematode that lived in the wood of the pine and was named the pine wood nematode. The pine wood nematode invaded the resin canals of the infected pines and caused a vascular disease similar to the wilt diseases caused by fungi (see Chapter 12, Wilt Diseases). Since the pine wood nematode also has been found recently in many areas in the United States, there has been great concern by plant pathologists that an epidemic of pine wilt similar to that in Japan would occur. An epidemic of the pine wilt disease has not occurred, however, and nematologists have since determined that the pine wood nematode was native to the United States and was introduced into Japan.

There is still debate among tree pathologists as to the importance and threat of the pine wood nematode to forest and shade tree conifers in the United States. Most agree that the importance of the pine wood nematode in the northern United States is minor, but its potential as a major tree pathogen in the southern United States is still uncertain. In any case the professional should be able to recognize and understand the pine wilt disease caused by the pine wood nematode.

Hosts The pine wood nematode (*Bursaphelenchus xylophilus*) has been found in the United States on a large number of pine species as well as on Atlantic and deodor cedar, European larch, tamarack, and white spruce (Wingfield *et al.*, 1982).

Symptoms Foliage color changes from green to light or lime green, yellow, and finally brown in rapid succession. Needles may drop during this time. The wood of an affected tree is noticeably dry and resin free. Death of affected trees occurs within a few months of the onset of symptoms.

Disease Cycle In the United States the pine wood nematode can exist as a nonpathogen in the wood of dead conifer trees or can cause the pine wilt disease. The pine wood nematode is carried by insect vectors—several species of pine sawyer or long-horned beetles. These insects commonly deposit their eggs in the wood of dead and dying conifers. During the pupation stage of the developing beetles, the larvae of the pine wood nematode, if present in the wood, are attracted to the beetle. The nematode larvae colonize the pine sawyer beetle and are carried with the adult beetles when they leave the tree. The adult beetles fly to the branches of live conifers to feed before mating and completing their life cycle.

It is at this stage that the pine wood nematode has the opportunity to become a pathogen and cause the pine wilt disease. The pine wood nematodes leave the body of the beetle during its feeding on live branches and enter the tree through the fresh feeding wounds. If the nematodes can successfully enter these wounds, multiply rapidly, and invade the resin canals of the tree, they can rapidly cause the complete collapse of the vascular system. In a matter of only a few weeks after this invasion the needles of the tree lose turgidity and the foliage turns yellow and then brown in rapid succession. Most affected trees die in one season.

In the United States, however, the pine wood nematode usually does not cause the pine wilt disease in trees, despite the feeding wounds of adult pine sawyer beetles. The pine wood nematodes are often carried to dead and dying trees, which are attractive to the adult female beetle for brood sites. The nematodes are deposited in these trees along with the insect eggs. The nematodes multiply in the dead or dying tree and colonize the next developing beetle brood to complete their life cycle.

The pine wood nematode can be easily isolated from dead or dying trees. Its role, therefore, in the death of a tree can often be confused. For example, many coniferous trees dying of root rot will be colonized by secondary insects, such as the pine sawyer. These trees often will yield pine wood nematodes upon sampling. In this case the diagnostician may mistake the role of the pine wood nematode as that of a pathogen, when decline and death in fact were caused by root rot fungi.

Treatment Affected trees should be removed as soon as they are detected since dying and dead coniferous trees serve as brood trees for the pine sawyers as well as numerous other important tree-attacking insects, such as bark beetles. All cut logs and branches should be taken off site and placed in a landfill. Since pine sawyers are attracted to weak trees initially, efforts to minimize stress and improve vigor (see Chapter 21, Tree Maintenance) will decrease chances of attack on specimen shade trees.

Foliar Nematodes

Some plant-parasitic nematodes can also attack the foliage of small trees and shrubs in subtropical and tropical regions. Due to the mild conditions and abundant moisture, these foliar nematodes can attack leaves, causing yellow and brown streaks to occur. Nematodes in the genus *Aphelenchoides* are most often associated with such foliar injury. In Florida, the rubber plant and azaleas are common woody plant hosts for foliar nematodes. Removal of affected leaves, sanitation of fallen leaves, and application of nematicides have proven effective in controlling foliar nematode problems.

LITERATURE CITED

Wingfield, M. J., R. A. Blanchette, T. H. Nichols, and K. Robbins. (1982). The pine wood nematode: A comparison of the situation in the United States and Japan. *Can. J. Forest Res.* **12,** 71–75.

SUGGESTED REFERENCES

Benson, D. M., J. T. Walker, and K. R. Barker. (1982). Controlling nematode damage. *Am. Nurseryman* **155,** 85–89.
Dropkin, V. H., and M. Linit. (1982). Pine wilt—A disease you should know. *J. Arbori.* **8,** 1–6.
Rohde, R. A. (1960). The nature of nematode damage to shade trees and recommended control practices. *Proc. Int. Shade Tree Conf.* **36,** 143–146.
Ruehle, J. L. (1964). Nematodes, the overlooked enemies of tree roots. *Prc. Int. Shade Tree Conf.* **40,** 60–67.
Ruehle, J. L. (1972). Nematodes and forest trees. *In* Economic nematology (J. M. Webster, ed.), Chapter 14, pp. 312–324. Academic Press, New York.
Ruehle, J. L. (1973). Nematodes and forest trees—Types of damage to tree roots. *Annu. Rev. Phytopathol.* **11,** 99–118.
Sluggett, L. J. (1972). Corky root disease of Douglas-fir nursery seedlings. *Can. For. Serv., Pest Leaf.* No. 53, 1–5.
Zuckerman, B. M., W. F. Mai, and M. B. Harrison (eds.). (1985). Plant nematology laboratory manual. Univ. Mass. Agric. Expt. Sta., Amherst, MA, 212p.

5

Viruses

INTRODUCTION

Viruses are "submicroscopic entities," that is, they are too small to be seen even with the aid of a powerful light microscope. They can only be seen with an electron microscope capable of magnifying them many thousands of times. All known viruses are parasitic on either plants or animals, including humans. These pathogens are known to cause a variety of diseases in people, including the common cold, flu, rabies, smallpox, yellow fever, and warts.

Viruses may occur in many shapes but most are either rod-shaped or polyhedral (Fig. 5.1). They are characteristically composed of a protein coat and a hollow nucleic acid core. The core may be composed of either DNA (deoxyribonucleic acid) or RNA (ribonucleic acid). Viruses have no cell wall, membranes, cytoplasm, organelles, or nucleus. They simply contain genetic material in the form of nucleic acid, for reproduction, which is protected by a protein coat. Viruses appear to be on the threshold of life, being more an organic chemical than a living organism but still being capable of causing diseases. In many cases, two or more components of a virus must combine to produce a disease.

HOW VIRUSES CAUSE TREE DISEASE

Viruses are parasites on plants and animals but, in contrast to the infectious pathogens discussed previously, viruses do not feed on cells or tissues. The only thing viruses do within the host is to reproduce (replicate) themselves, and since viruses have such simple bodily structures they do not have the means to replicate unless they are inside a host cell. Viruses carry the instructions for the production of new viruses in their nucleic acid core. Once inside the cell the viral protein coat is shed and the virus particle causes the host cell to manufacture viruses identical to itself. It also causes the cell to neglect its necessary metabolic functions such as production of chlorophyll. The net result is that the cell

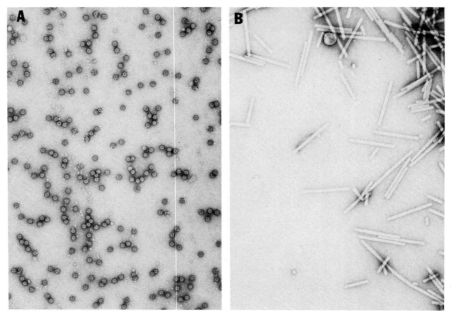

Fig. 5.1 (A) Polyhedral virus, cowpea chlorotic mottle virus (CCMV) (80,000×). (B) Rod-shaped virus, tobacco mosaic virus (TMV) (83,000×). (Photos courtesy of George N. Agrios, University of Massachusetts, Amherst.)

reproduces the virus but is unable to function normally. The infected host cell may either grow and divide rapidly or it may grow very slowly and be unable to divide, but the end result of virus infection in either case is the abnormal growth and function of infected cells.

Viruses, once inside a host organism, are able to constantly infect new cells. Viruses are termed systemic pathogens because they eventually spread internally throughout the infected host plant. In the case of a large tree it could be 1 to 4 years before a virus could spread throughout the plant. The effect of viruses on one cell is compounded by the combined effects of the other infected cells around it. The total effect of many virus-infected cells can be seen as some type of abnormality (symptom) on some part(s) of the plant such as leaves, flowers, shoots, twigs, or small roots. It would, therefore, seem logical that soon after a large part of a tree became symptomatic of a virus disease it would die, but most virus-infected trees usually live for many years after infection. Since viruses are obligate parasites they need young active cells for replication. It is the tissues that are newly produced, each year in the spring by the buds, at the cambium, and in the roots, that are attacked by the virus, but fortunately for the tree not all the tissues are infected each spring. Once the tissues slow in growth in the summer and mature, they become resistant to virus infection. Trees and shrubs are often able to survive virus infection for long periods because they alternate periods of rapid cell growth, little cell growth, and dormancy. Virus-infected trees, therefore, usually prevent complete infection of all cells by producing a complete set of new cells and organs each growing season that must be reinfected by the virus.

Most viruses are spread from infected trees to healthy trees by two methods: (1) insect transmission, and (2) wounding. Insect transmission occurs primarily as a result of feeding of insects with sucking mouth parts, such as aphids and leafhoppers (Fig. 5.2). These insects feed by inserting daggerlike mouth parts (stylets) into the vascular tissue of a leaf and withdrawing sap. The sucking insects first feed on the sap of a virus-infected tree or shrub and then may move to a healthy individual. By the processes of inserting their stylets into the leaf and feeding, virus particles from the insect are injected into the healthy plant, and the process of infection begins. Wounding or mechanical transmission sometimes occurs during vegetative propagation or pruning. If virus-infected propagating material is budded or grafted onto a healthy tree or onto another rootstock the outcome will be a virus-infected tree in almost every case. If a

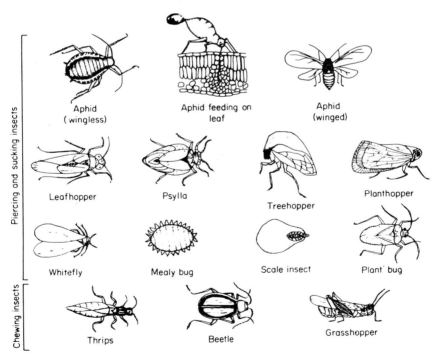

Fig. 5.2 Insect vectors of plant viruses. Insects in second row also transmit mycoplasma-type pathogens. [Drawing courtesy of George N. Agrios. (1977). Plant pathology. Academic Press, New York.]

healthy tree is pruned with the same pruning tools (especially hand tools), without disinfection, that had previously been used on a virus-infected tree, the virus may be transmitted to the healthy tree. Viruses may also be spread by nematodes during feeding, in a manner analogous to insect transmission, but this means of spread is rare. Other infrequent means of transmission are viruses carried in the seed, or in some cases, transmitted by the pollen of an infected plant during fertilization of the flower.

SYMPTOMS

The symptoms caused by each virus disease of a tree or shrub is unique. A virus disease may cause distinctive symptoms, quite subtle symptoms, or it may produce no symptoms at all. The symptoms of most virus diseases can be put into four categories for ease of identification. Most virus diseases of woody plants involve one or more of the following general symptoms: (1) lack of chlorophyll formation, in normally green organs, (2) dwarfing or growth inhibition, (3) distorted growth of part of or the entire plant, and (4) necrotic areas or lesions.

Chlorophyll formation may partially be lacking, resulting in light green areas, or may be severely inhibited, resulting in areas appearing yellow or white. The net result of alternate areas of normal chlorophyll formation and areas of inhibited chlorophyll formation is a pattern of light and dark green color that can take a variety of forms (Fig. 5.3). If the affected areas cover relatively large areas and appear in a random patchwork fashion, the symptom is referred to as a mosaic or a mottle. If the affected tissue occurs in open circles or ellipses, the pattern is called a ring spot, and it is called a blotch when the circles or ellipses are closed. A linear form of this symptom is referred to as a line pattern (Fig. 5.4).

The virus-infected tree or shrub may be smaller than healthy individuals of the same age, because of the detrimental effects of the virus on the utilization of sunlight and subsequent formation of energy reserves. This may be difficult to diagnose, however, especially in trees where the suspected one is not the same age as others of its species, or if it happens to be the only one of its species that was planted at a particular time. In addition, virus-infected woody plants are often sterile while the healthy individuals of the same species and comparable size produce flowers and seed.

Viruses may affect the normal growth pattern of woody plants resulting in a variety of formations on plant parts that are atypical. Rapid division and growth of the virus-infected cells often occurs. Affected leaves may twist and curl or become blistered or puckered in some areas. Large numbers of shoots may appear from single branches or the trunk and form witches' brooms. Branches may be flattened (Fig. 5.5), and in some other cases they are unable to support themselves and become limp or rubbery.

Necrotic areas may appear on any part of a virus-infected plant. Local leaf

Some Common Virus Diseases

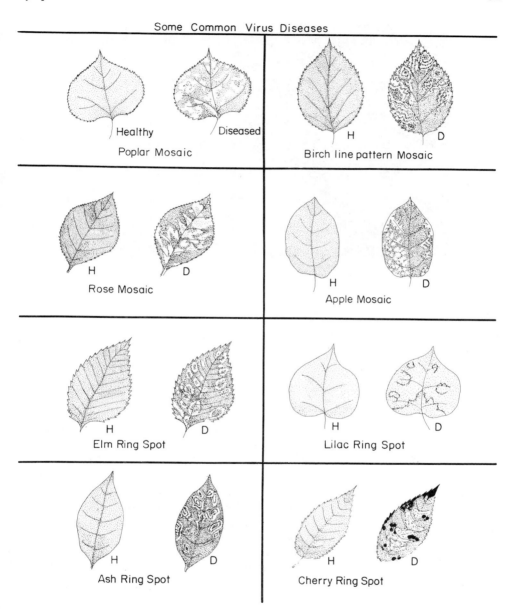

Fig. 5.3 Some common virus diseases of woody plants.

Fig. 5.4 Line pattern on white ash caused by TMV. (Photo courtesy of George N. Agrios, University of Massachusetts, Amherst.)

lesions that appear as necrotic spots or rings are the most common symptom but necrotic streaks under the bark of branches may also occur.

TREATMENT

Most control efforts with virus diseases of woody plants center on obtaining and maintaining virus-free plants. For the most part the key to virus control lies with the nurserymen. Since most viruses are easily transmitted through vegetative propagation, the utmost care must be employed when selecting virus-free source material for budding and grafting. The "mother" trees should be checked periodically (indexed) at the nursery for the presence of viruses. These trees should also be protected from insect vectors by insecticide sprays. Grafting tools should be surface-sterilized between trees to prevent any chances of transmission from the sap. The arborist, municipal tree official, or homeowner, therefore, should attempt to obtain virus-free plants for use as shade and ornamental trees by dealing only with reputable nursery operations that are known for the production of high quality planting stock.

Viruses in a woody plant can sometimes be inactivated by heat treatment. This is an exacting procedure that usually involves the total immersion of small trees or shrubs in water just hot enough to inactivate the virus [110°F (43°C) to 135°F (57°C)] but not hot enough to kill the plant. The amount of time required for inactivation varies with the virus and the temperature used. In many cases the plants are killed from the treatment. This type of therapy is used primarily in

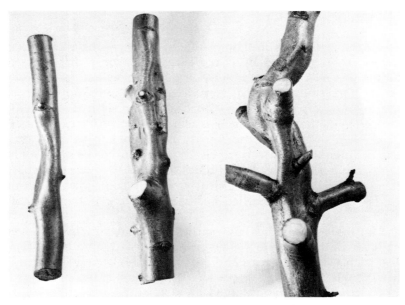

Fig. 5.5 Flat limb disease on apple. (Photo courtesy of George N. Agrios, University of Massachusetts, Amherst.)

nurseries to obtain virus-free plants for propagation and is not widely used in arboriculture.

A virus-infected shade tree will often live for many years after infection. It may go completely unnoticed by the average homeowner and often by all but the trained plant pathologist. Although virus-infected trees grow slower than virus-free trees, once the tree has grown large enough to serve its function for shade and beauty it is often an attractive and useful specimen despite a virus infection. In these cases, virus-infected trees, therefore, need not be removed. They should be given proper tree maintenance including watering when needed and fertilizing each year, due to their low vigor and susceptibility to other diseases and environmental stresses. In other cases, where the virus disease results in severe distortions in the tree's normal form, it is no longer functional as a shade or ornamental tree and should be removed. Adjacent trees, especially those of the same species, may be protected against insect transmission of the virus by insecticide sprays. However, this method of protection is often unsuccessful since insect-borne viruses are often inoculated into the tree through feeding before the insect is killed by the insecticide. Young seedling trees that are infected with viruses, regardless of type, have little chance to become functional shade or ornamental trees and should be removed and destroyed as soon as detected.

VIRUS DISEASES OF TREES

Viruses are not given Latin binomials like other organisms but instead are described by symptoms and host. For example, a virus that produces chlorophyll inhibition in a random pattern on poplar trees is referred to as poplar mosaic virus. Occasionally a particular virus that is well known for causing a virus disease on one plant, such as tobacco mosaic virus (TMV), will cause a virus disease on another species such as ash. When the true identity of the virus on ash has been determined the disease is referred to as TMV on ash. To determine the identity of a virus, however, usually involves inoculations of the virus into many herbaceous plants (indicator plants) known to display typical symptoms when infected with a particular virus. In some cases, electron micrographs of the virus are also needed for the identification of an unknown virus disease.

New virus diseases of trees and shrubs are being discovered every year. However, most of the known virus diseases of woody plants are those found on fruit trees, such as apples, citrus, and peaches, and relatively few are known on shade trees (Table 5.1). One explanation is that most of the tree virus research has been on fruit trees with only small amounts of research on forest and shade trees. Since many known plant viruses attack several species of woody and nonwoody plants, it is likely that a larger number of virus diseases of shade and ornamental woody plants will be discovered as more research is conducted.

TABLE 5.1

Some of the Common Virus Diseases of Shade Trees in the United States

Host	Disease
Ash	Ash ringspot (tobacco ringspot virus)
	Ash line pattern (tobacco mosaic virus)
Birch	Birch line pattern (apple mosaic virus on birch)
Black locust	Black locust witches' broom (tomato spotted wilt virus)
Edler	Tobacco ringspot virus on elder
Elm	Elm mosaic virus
	Elm ringspot virus
	Elm scorch virus
	Elm zonate canker virus
Lilac	Lilac ringspot virus
Maple	Maple mosaic (maple mottle)
	Peach rosette virus on maple
	Tobacco ringspot virus on maple
Oak	Oak ringspot
Poplar	Poplar mosaic virus

TABLE 5.2

Some of the Common Virus Diseases of Flowering Apples, Cherries, Peaches, Plums, and Pears

Host	Disease
Apple	Apple mosaic
	Chlorotic leaf spot
	Stem pitting
	Flat limb
Cherry	Stem pitting of Prunus (tomato ring spot virus)
	Line pattern
	Prunus necrotic ringspot
	Raspleaf
Pear	Stony pit
	Vein yellows and red mottle
	Ring pattern mosaic
Plum	Stem pitting of Prunus (tomato ringspot virus)
	Sour cherry yellows (prune dwarf virus and prunus necrotic ringspot virus)
Peach	Ringspot
	Yellowbud mosaic (stem pitting of prunus) (tomato ringspot virus)
	Mosaic

VIRUS DISEASES OF ORNAMENTAL APPLES, CHERRIES, PEACHES, PEARS, AND PLUMS

Virus diseases on commercial fruit trees have been known for many years to cause serious crop losses, and most of what is known about tree viruses is based upon research with fruit trees. Most of these same virus diseases can also occur on the closely related ornamental and flowering varieties of these fruit trees (Table 5.2). Careful screening to eliminate virus-susceptible varieties is employed during propagation of commercial fruit tree varieties, in an attempt to minimize losses from virus diseases. Ornamental varieties, however, are selected primarily for aesthetic properties, such as form and flowering, and are usually not indexed for virus susceptibility. In fact, the susceptibility of certain ornamental varieties such as Nanking cherry and crab apples to common fruit tree viruses is so high that these trees are often used as indicator plants for virus detection in commercial fruit varieties. However, the susceptibility of most of these ornamental tree varieties, especially those most recently developed, to most of the common commercial fruit virus diseases is still unknown. Ornamental fruit trees appear, in general, to be the most likely group of trees to have virus diseases. These tree varieties are becoming increasingly popular for shade trees, and these varieties should be watched most carefully for symptoms of virus diseases.

SUGGESTED REFERENCES

Agrios, G. N. (1975). Virus and mycoplasma diseases of shade and ornamental trees. *J. Arbori.* **1,** 41–47.

Anonymous. (1951). Virus diseases and other disorders with virus like symptoms of stone fruits in North America. *U.S., Dep. Agric., Agric. Handb.* **10,** 1–276.

Matthews, R. E. F. (1981). Plant virology, 2nd ed. Academic Press, New York, 897p.

Posnette, A. F. (ed.). (1963). Virus diseases of apples and pears. Commonw. Agric. Bur. Tech. Commun. No. 30. Farnham Royal, Bucks, England.

6

Mycoplasmas

INTRODUCTION

All virus and mycoplasma diseases of plants were thought to be caused by viruses, until recently. It is now known, however, that there are many differences between these two types of pathogens, not only in their structure, but also in the way they attack plants and in the symptoms they cause. Mycoplasmas are much larger than viruses and some stages of them are even smaller than bacteria although mycoplasmas are often longer than bacteria. The mycoplasmas that attack plants are usually spherical or occur in long filaments. In contrast to viruses, which resemble organic chemicals, mycoplasmas do not have a cell wall, but like the bacteria, they contain a plasma membrane, cytoplasm, and strands of DNA. Their nondistinctive appearance and small size could be two reasons why the mycoplasmas associated with plant diseases were overlooked until 1967 and thought to be viruses. It is, therefore, still quite difficult to identify mycoplasmas as the causal agent of a plant disease, and many more mycoplasma diseases probably exist than the few that are known presently.

The mycoplasmas are parasites of plants and animals or they are saprophytes. Once in a living cell they may divide by fission, in a manner similar to bacteria; by budding, where a small swelling forms on the outside of the cell and is eventually pinched off to form a new mycoplasma; or through constrictions along the body in the filamentous forms, which break off and produce several additional mycoplasmas.

HOW MYCOPLASMAS CAUSE TREE DISEASE

The exact mechanisms by which mycoplasmas cause tree diseases are unknown, but the effects of the pathogen on the host give us some clues. Mycoplasmas, like bacteria, have very limited mobility. Mycoplasmas, however, are able to effectively utilize the vascular system to move throughout the host plant;

therefore, most mycoplasma diseases are usually considered systemic diseases. Large numbers of mycoplasmas are often found in the phloem. They occur most frequently in the sieve elements of this tissue, and may interfere with the movement of sugars through the plant. Blockage of the movement of these essential energy-storage compounds from leaves to roots could account for the progressive decline and death often associated with mycoplasma diseases. These diseases, however, are often associated with growth abnormalities and necrosis on the host as well as apparent starvation symptoms. It may be a combination of vascular blockage, energy lost to mycoplasma reproduction, and induced growth abnormalities that cause mycoplasma diseases in trees.

Mycoplasmas can be spread from tree to tree in some of the same ways that viruses are spread: vegetative propagation, insect vectors, and root grafts. Insect transmission appears to be the most important mechanism for spread of mycoplasma diseases of trees. Leafhoppers are the most common vectors, with plant hoppers and psyllids also being vectors of some mycoplasmas (Fig. 5.2). In contrast to most virus diseases that are insect transmitted, mycoplasmas usually incubate 10 to 20 days within the insect's body before the pathogen can be inoculated into a new host. Although this phenomenon is also known for some viruses, it is not common, and most viruses can be transmitted to several new hosts in the same day. When the mycoplasmas are picked up by the stylet of the insect during feeding they appear to infect the insect and multiply through its body. This insect–mycoplasma association is often lifelong for the insect, and in some cases the mycoplasmas are even passed from the adult to offspring in the eggs. Hence, both the adult and its offspring can inoculate mycoplasmas into plants throughout their lives.

SYMPTOMS

The universal symptom of mycoplasma diseases appears to be a progressive decline that eventually results in death of the host in a few months to several years. The decline may be associated with any or all of the following symptoms: chlorotic foliage, little terminal growth or stunting, progressive dieback of branches, lack of flowering, and necrosis of phloem tissues. Other common symptoms of mycoplasma diseases are reddening of foliage, witches' brooms on branches and on axillary shoots, and proliferation of roots.

TREATMENT

Since a mycoplasma, like a virus, moves throughout the tree after infection, attempts at control must be made either by prevention of infection or by therapeutic treatment of the host. Mycoplasma diseases can usually be prevented in the same manner as virus diseases since they share many common modes of transmission. Care should be exercised during vegetative propagation to ensure only healthy grafting materials are used.

Mycoplasmas as a group are sensitive to tetracycline antibiotics, which include tetracycline, oxytetracycline, Terramycin, Achromycin, Aureomycin, and demethychlor tetracycline. Since antibiotic sensitivity is unknown in viruses, this sensitivity was an early clue that some of the diseases thought to be caused by viruses might be caused by some other pathogens. Tetracycline antibiotics have been used with various degrees of success to control many mycoplasma diseases of plants. Treatment of mycoplasma diseases of trees has been most successful when the antibiotics were injected directly into the wood. Antibiotic treatments, however, result in remission of symptoms for a few years at best and do not rid the host of the mycoplasma.

MYCOPLASMA DISEASES OF TREES

Elm Phloem Necrosis or Elm Yellows

Hosts American, cedar, red (slippery), rock, and September elm are known to be susceptible. European and Asiatic elms and their hybrids are resistant to phloem necrosis.

Symptoms The first aboveground symptoms are the appearance of a combination of leaf yellowing (chlorosis), leaf droop (epinasty), leaf curl, and premature leaf drop during mid to late summer (Fig. 6.1). In red (slippery) elms, witches' brooms are often produced (Fig. 6.2). These symptoms usually occur over the entire crown and progressively increase in severity to give the tree a "starved" appearance. In some cases, however, these initial symptoms may develop in only one branch before they occur in the rest of the crown. Trees with severe decline symptoms in the fall often do not produce any leaves in the spring. Infected trees may also produce small leaves or completely normal-appearing leaves in the spring. The dwarfed leaves usually turn yellow and fall shortly after formation. In late spring the foliage on fully leaved, infected trees usually wilts and dies rapidly, but remains attached to the tree. In general, death of the tree occurs within 1 year following initial symptom expression and natural recovery has not been detected. Occasionally, however, a tree may survive several years after infection.

The first symptoms of phloem necrosis can be found in the roots. Small feeder roots are killed initially and then larger roots are progressively killed. The killing occurs primarily in the active phloem tissue next to the cambium (inner bark) and results in a progressive yellow or "butterscotch" discoloration of this area. The discoloration can most easily be seen in the lower trunk and root buttress area, although the pathogen usually causes some discoloration of the inner phloem throughout the tree during the course of the disease (Fig. 6.3). The discoloration associated with phloem necrosis does not occur more than super-

Fig. 6.1 American elm branches, healthy (left) and infected with phloem necrosis (right). Note light leaf color, and leaf droop in the infected branch. (Photo courtesy of Wayne A. Sinclair, Cornell University, Ithaca, New York.)

Fig. 6.2 Witches' brooms in red (slippery) elms infected with phloem necrosis (A) summer, (B) winter. (Photos courtesy of Wayne A. Sinclair, Cornell University, Ithaca, New York.)

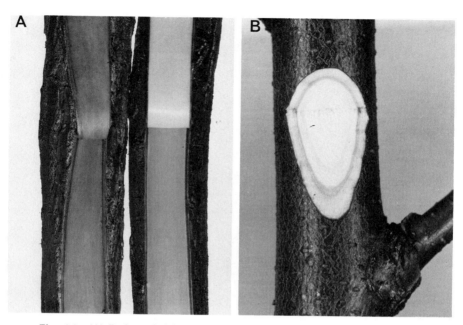

Fig. 6.3 (A) Bark peeled back to expose inner phloem and cambium region of healthy (left) and phloem necrosis-infected American elm stems. Note the discoloration in the inner phloem and cambium region of the infected stem. (B) Section exposing both xylem and phloem of a phloem necrosis-infected stem. Note discoloration in the inner bark–cambium region but lack of discoloration in the xylem. (Photos courtesy of Wayne A. Sinclair, Cornell University, Ithaca, New York.)

ficially (1 mm or less) in the xylem and usually can be scraped off easily with a knife. Discoloration deeper in the xylem is usually associated with other vascular diseases, principally Dutch elm disease and Verticillium wilt, which can sometimes be confused with elm phloem necrosis (see Chapter 12, Wilt Diseases). Occasionally, these vascular diseases, usually Dutch elm disease, will occur simultaneously with phloem necrosis.

The most useful symptom for detection of elm phloem necrosis is the oil of wintergreen (methyl salicylate) odor that is produced in the phloem of infected American, cedar, and September elms. This common odor is familiar to most people from its many uses in foods, medicines, and deodorizers, and occurs only in elms that are infected with the phloem necrosis mycoplasma. The amount of this chemical in the phloem, however, is often quite small and the odor from freshly exposed inner bark may be difficult to detect. A sample of this tissue should be placed in a closed container (plastic bag, small jar) for a few minutes. Periodically sniff the container for the characteristic odor. The oil of wintergreen odor is usually not easy to detect in the winter. When detection is attempted during the dormant season, the container with bark sample should be

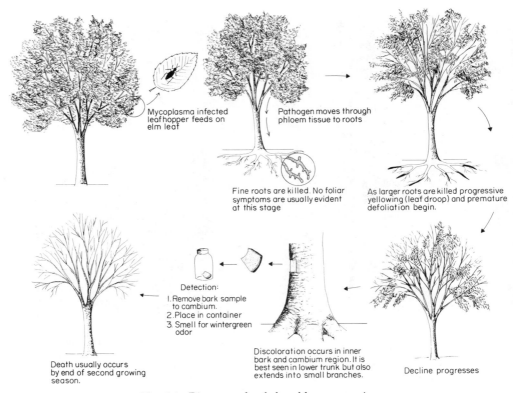

Fine roots are killed. No foliar symptoms are usually evident at this stage

Pathogen moves through phloem tissue to roots

Mycoplasma infected leafhopper feeds on elm leaf

As larger roots are killed progressive yellowing (leaf droop) and premature defoliation begin.

Detection:
1. Remove bark sample to cambium.
2. Place in container
3. Smell for wintergreen odor

Discoloration occurs in inner bark and cambium region. It is best seen in lower trunk but also extends into small branches.

Death usually occurs by end of second growing season.

Decline progresses

Fig. 6.4　Disease cycle of elm phloem necrosis.

warmed by body heat or taken indoors and incubated for much longer periods. A "maple syrup" odor is often produced by the inner bark of red (slippery) elms shortly after the bark is killed. No odor is produced as long as the bark is still alive.

Disease Cycle　The most common cause of infection is through the leaf feeding of a mycoplasma-infected insect (Fig. 6.4). The white-banded American leafhopper is the only known vector of this pathogen. The disease may be limited in its range due to the natural range of the insect vector. The disease is not found in areas where the average annual minimum temperatures are below −15°F (−26°C). Intolerance to extreme cold of the insect vector could explain this northern limitation of the disease. Infection may also occur through root grafts from an infected tree.

The pathogen, once inside the host, moves through the phloem throughout the tree including the root system. The first necrotic symptoms occur in the small roots, which are killed; progressively larger roots then are killed. Discoloration in the inner bark occurs in the buttress area and eventually throughout the tree. Foliar symptoms usually begin at this time with yellowing, leaf droop, leaf curl, and premature defoliation. The tree progressively declines as more and more of the root system is killed, and phloem transport is blocked. If infection has occurred early in the growing season the tree may not produce leaves the next spring. Death usually occurs during the second growing season.

Treatment   Outbreaks of phloem necrosis can be prevented by prompt removal of infected hosts. Root grafts should also be severed between infected and healthy elms. Recent experiments with a tetracycline antibiotic, oxytetracycline, that was trunk injected into American elms already infected with phloem necrosis, have resulted in symptom remission for 3 years. Tetracycline antibiotics are currently being used to effectively control X-disease on peach, a damaging mycoplasma disease of commercial peach and cherry orchards, and similar control techniques for elm phloem necrosis and other mycoplasma diseases of shade trees may soon be developed. European and Asiatic elms planted in areas where phloem necrosis occurs, appear to possess some degree of resistance. Since these trees and their hybrids are currently being used for replacement of American elms that have been killed by Dutch elm disease they are perhaps also the best choices of elm species in areas where phloem necrosis is present.

Coconut Lethal Yellows

Hosts   Coconut palms are the most commonly infected palms but many other species of palms, including the Christmas palm and several date palms, are also susceptible. A number of resistant coconut palm varieties have been developed. The Malayan dwarf variety is the most widely used but other varieties, including the Maypan, King, Ceylon Dwarf, Red Spicata, and Fiji Dwarf palms have been developed.

Symptoms   Initial symptoms usually are dropping of coconuts followed shortly by blackening and death of flower parts. Leaves begin to turn yellow and progressively die up the tree. Brown streaks can usually be seen at the tips of the youngest leaflets before they are killed. The dead brown leaves often remain attached to the tree for some time. Eventually, however, all the leaves die and fall from the trunk giving it a "telephone pole" appearance.

Disease Cycle   It is not known exactly how the mycoplasmas that cause this disease enter the plant, but it is suspected that an insect vector is involved (Fig. 6.5). The pathogen, once in the host, appears to quickly cause the death of floral parts followed by the leaves. Death of the trees usually occurs from 3 to 6 months after the first symptom appearance.

Treatment   Early detection and eradication of diseased trees will minimize local spread of the pathogen and limit outbreaks. Since the Malayan dwarf variety of coconut palms appears to be highly resistant or immune to lethal yellows disease it should be selected for ornamental plantings. Palm species that are not affected by lethal yellows should also be considered. The sabal palmetto or cabbage palm and the royal palm are two examples of palms that are not affected by lethal yellowing.

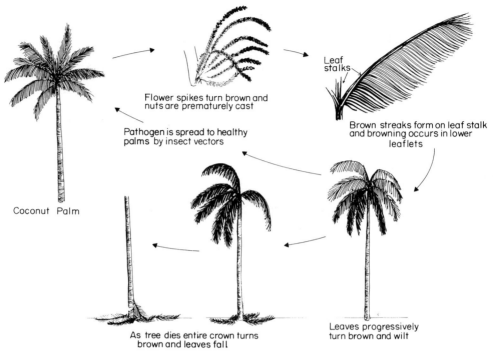

Flower spikes turn brown and
nuts are prematurely cast

Pathogen is spread to healthy
palms by insect vectors

Leaf
stalks

Brown streaks form on leaf stalk
and browning occurs in lower
leaflets

Coconut Palm

Leaves progressively
turn brown and wilt

As tree dies entire crown turns
brown and leaves fall

Fig. 6.5 Disease cycle of lethal yellows of coconut palms.

Both preventative and therapeutic treatment of trees with oxytetracycline anti-biotic has been somewhat successful in controlling this disease. The antibiotic is injected into the trunk either by standard gravity or pressure injection tech-niques. As with most diseases regardless of pathogen, therapeutic treatment is most successful in the earliest stages of the disease. Once one fourth or more of the leaves have yellowed, antibiotic treatment will, in most cases, be unsuc-cessful, and the trees should be removed. Although therapy may cause remis-sion of symptoms the antibiotic does not cure the tree of the lethal yellowing disease. The treatments must be repeated every 4 months for the life of the tree or symptoms will reappear and death will occur.

Ash Yellows

Ash yellows, which is sometimes called ash yellows and witches' broom or ash witches' broom, is a mycoplasma disease that is a component of the complex disease called ash decline (see Chapter 23, Diebacks and Declines—Complex Diseases). In recent years many white and green ash trees, both in forests and in shade tree locations, have exhibited the symptoms of ash decline. This decline has resulted in substantial losses in both forest and shade ash trees. The role of the mycoplasma disease ash yellows in ash decline and subsequent mortality is a subject of concern and debate among tree pathologists. Some pathologists have shown a strong relationship between ash yellows and ash decline throughout the northeastern and midwestern United States. Other pathologists have found a close correlation between moisture stress and ash decline. Some studies have presented ash mycoplasma as a primary cause of ash decline, while others saw it as a secondary problem that resulted originally from environmental stresses. No one, however, debates the importance of the widespread mortality of ash species.

Hosts White and green ash appear to be the principal hosts, although white ash is the most severely affected species. In addition, the widely used green ash cultivar "Marshall seedless" was found to be infected and displayed witches' brooms (Carr 1986).

Disease Cycle A leafhopper vector is suspect. It has not been determined how long it takes after infection occurs in an ash tree for symptom onset to occur. Most healthy-appearing green and white ash shade trees in this study tested positive for the presence of the ash yellows mycoplasma (Carr, 1985). Infected trees, however, often display a history of at least several years of poor incremental growth before visible symptoms appear. Slow twig growth at the branch terminals results in tufted foliage and, later, in loss of terminal dominance. Foliage may be a chlorotic yellow or yellow-green. Progressive twig and branch dieback then occurs, usually over several seasons. Sprouting may occur on major limbs, the trunk, and around the root collar. These sprouts may later become abnormal in form and turn into witches' brooms, in the later stages of this disease when most of the crown has died. Severe bark seams may also occur in the trunk at this stage of the disease. Death usually follows within a year of severe crown dieback.

Treatment The ash yellows mycoplasma appears to be widespread among ash trees in both forest and shade tree sites. Ash trees free of ash yellow symptoms should be kept as vigorous as possible and protected from stress. Regular fertilization, watering, and mulching are recommended. Severely infected ash trees should be removed to prevent insect transmission of nearby nonsymptomatic trees. The use of ash species in ornamental plantings should be discouraged.

OTHER COMMON MYCOPLASMA DISEASES OF SHADE TREES

Most of the mycoplasma diseases of trees are associated with witches' brooms on the trunk and branches and the progressive decline of affected trees. Table 6.1 lists some additional common mycoplasma diseases of shade trees and their hosts. Although insect vectors are suspected in these diseases, none are currently known.

TABLE 6.1

Some Common Mycoplasm Diseases of Shade Trees

Hosts	Disease
Black locust	Black locust witches' broom
Hickory (native species)	Pecan bunch disease
Pecan	Pecan bunch disease
Black walnut	Walnut bunch disease
Butternut	Walnut bunch disease
English walnut	Walnut bunch disease
Japanese walnut	Walnut bunch disease
Willow	Willow witches' broom

MYCOPLASMA DISEASES OF ORNAMENTAL APPLES, CHERRIES, PEACHES, AND PEARS

Mycoplasma diseases of commercial fruit trees were discovered soon after the first mycoplasma diseases of plants were found in 1967. Most of the mycoplasma diseases were well known but all were originally thought to be caused by viruses. Our knowledge of mycoplasma diseases of trees, like our knowledge of virus diseases of trees, is built to a large extent upon research experience with mycoplasma diseases of commercial fruit trees. Some of the same problems that are present with ornamental varieties of commercial fruit trees, such as lack of screening and indexing for virus diseases, also exist for the common mycoplasma diseases. Also, as in virus disease, the susceptibility of ornamental tree varieties to the common mycoplasma diseases is still unknown.

Table 6.2 lists some of the common mycoplasma diseases of flowering apples, cherries, peaches, and pears.

TABLE 6.2

Some Common Mycoplasm Diseases of Flowering Apples, Cherries, Peaches, and Pears

Hosts	Disease
Apple	Proliferation disease
Cherry	X-disease
Cherry	Rusty mottle
Peach	Yellows
Peach	Peach rosette
Peach	Little peach
Peach	X-disease
Pear	Pear decline

CONFUSION OF VIRUS AND MYCOPLASMA SYMPTOMS WITH NONINFECTIOUS DISEASES

Both virus and mycoplasma diseases may sometimes be confused with injury from weed-killers (herbicides), air pollution injury, mineral deficiencies, or other noninfectious diseases (see sections on these areas). Since exact diagnosis of a virus or mycoplasma often requires extensive testing, it is best to make sure that these noninfectious problems can be eliminated as possible causes of the problem before making a diagnosis for a virus or mycoplasma disease.

LITERATURE CITED

Carr, K. P. (1986). Symptoms and distribution of ash yellows in Massachusetts and effects of oxytetracycline microinjection on Mycoplasma-like organisms within white ash. M.S. Thesis, Univ. of Mass., Amherst, MA, 74p.

SUGGESTED REFERENCES

Agrios, G. N. (1975). Virus and mycoplasma diseases of shade and ornamental trees. *J. Arbori.* **1**, 41–47.

Bloomfield, H. (1985). Lethal yellowing. *Am. For.* **91**, 38–40, 51–52.

Carter, J. C., and L. R. Carter. (1974). An urban epiphytotic of phloem necrosis and Dutch elm disease, 1944–1972. *Ill. Natl. Hist. Surv., Bull.* **31**, 113–143.

Filer, T. H., Jr. (1973). Suppression of elm phloem necrosis symptoms with tetracycline antibiotics. *Plant Dis. Rep.* **57**, 341–343.

George, J. C. (1976). The battle to save Florida's palms. *Natl. Wildl.* **14**, 17–19.

Granett, A. L., and R. M. Gilmer. (1971). Mycoplasmas associated with X-disease in various Prunus species. *Phytopathology* **61**, 1036–1037.

Hibben, C. R., and S. S. Silverborg. (1978). Severity and causes of ash dieback. *J. Arbori.* **4**, 274–279.

McCoy, R. E. (1974). How to treat your palm with antibiotic. *Fl., Agric. Exp. Stn., Circ.* **S-228**, 1–7.

McCoy, R. E. (1975). Oxytetracycline dose and stage of disease development on remission of lethal yellowing in coconut palm. *Plant Dis. Rep.* **59**, 717–720.

McCoy, R. E. (1982). Use of tetracycline antibiotics to control yellows diseases. *Plant Dis.* **66**, 539–542.

Maramorosch, K., and S. P. Raychaudhuri (eds.). (1981). Mycoplasma diseases of trees and shrubs. Academic Press, New York, 362p.

Martyn, R. D., and J. T. Midcap. (1975). History, spread, and other palm hosts of lethal yellowing. *Fl., Co-op. Ext. Serv., Circ.* **405**.

Matteoni, J. A., and W. A. Sinclair. (1982). Understanding ash decline. *Cornell Plantations* **38**, 42–45.

Matteoni, J. A., and W. A. Sinclair. (1985). Role of the mycoplasmal disease, ash yellows, in decline of white ash in New York State. *Phytopathology* **75**, 355–360.

Mullin, R. S., and D. A. Roberts. (1973). Lethal yellowing of coconut palms. *Fl., Co-op. Ext. Serv., Cir.* **354**, 1–4.

Nyland, G., and W. J. Moller. (1973). Control of pear decline with a tetracycline. *Plant Dis. Rep.* **57**, 634–637.

Sinclair, W. A., and T. H. Filer, Jr. (1974). Diagnostic features of elm phloem necrosis. *Arborist's News* **39**, 145–149.

Sinclair, W. A., E. J. Braun, and A. O. Larsen. (1976). Update on phloem necrosis of elms. *J. Arbori.* **2**, 106–113.

Wilson, C. L., and C. E. Seliskar. (1976). Mycoplasma-associated diseases of trees. *J. Arbori.* **2**, 6–12.

7

Seed Plants

Seed plants or higher plants, like most categories of microorganisms, are also capable of causing diseases of trees. These pathogenic higher plants are capable of acting as parasites directly on a tree or acting as effective competitors with trees for growing space and sunlight. The nonparasitic seed plants that cause tree disease are usually constricting or rapidly growing vines, which either strangle and/or shade out trees by growing completely over them. These vines can be troublesome shade tree problems because they are particularly aggressive near the roadside and in the backyard where the additional sunlight allows them to grow much faster than in the deeply shaded forest.

PARASITIC SEED PLANTS

These plants lack a normal root system and rely upon a host plant to supply them with the water and nutrients from the soil that they need to live. Parasitic plants, however, produce a specialized root system composed of structures called haustoria that are specifically adapted for absorbing water and nutrients from a host plant. Although there are a large number of species of parasitic seed plants, the most common problems on shade trees are caused by the dwarf mistletoes and the leafy mistletoes.

Dwarf Mistletoes

Dwarf mistletoes are plants that lack true leaves and contain only primitive leaves called bracts. These structures are unable to manufacture sufficient food via photosynthesis to meet the energy needs of the dwarf mistletoe plant, and most of its energy needs as well as water and minerals are supplied by the host tree. Because of this heavy reliance on the host tree, the dwarf mistletoes cause a great deal of damage to both forest and shade trees. Much of the trees' reserves

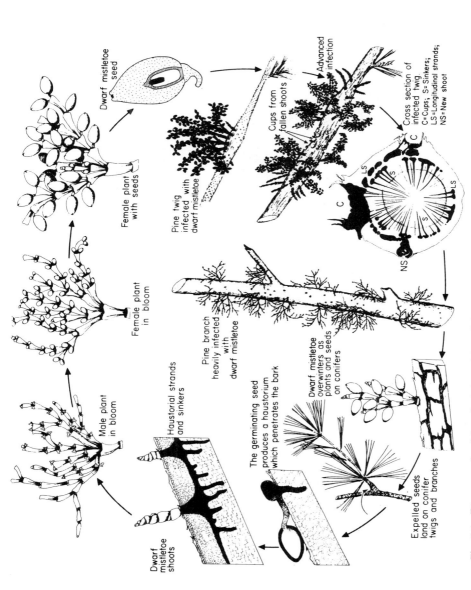

Fig. 7.1 Disease cycle of dwarf mistletoe (*Arceuthobium* sp.) on conifers. [Drawing courtesy of George N. Agrios. (1969). Plant pathology. Academic Press, New York.]

Dwarf mistletoe seed

Female plant with seeds

Female plant in bloom

Male plant in bloom

Dwarf mistletoe shoots

Haustorial strands and sinkers

Pine twig infected with dwarf mistletoe

Cups from fallen shoots

Advanced infection

Cross section of infected twig.
C=Cups; S=Sinkers;
LS=Longitudinal strands;
NS=New shoot

C

LS

S

LS

NS

Pine branch heavily infected with dwarf mistletoe

The germinating seed produces a haustorium which penetrates the bark

Dwarf mistletoe overwinters as plants and seeds on conifers

Expelled seeds land on conifer twigs and branches

Fig. 7.2   Male flowers of dwarf mistletoe. (Photo courtesy of USDA Forest Service.)

are diverted by the infecting dwarf mistletoe plants, and the host trees slow in growth and eventually decline and die.

Hosts   Dwarf mistletoes are restricted to coniferous hosts but many conifers in North America are susceptible to at least one species of dwarf mistletoe. As a group they are most prevalent in the western United States, western Canada, and Mexico and cause problems on ponderosa pine, lodgepole pine, Douglas fir, true firs, western larch, spruces and many other native conifers. In the north-eastern United States and eastern Canada black spruce is the most common host although white spruce is occasionally infected.

Symptoms   Infected branches increase in thickness. When compared at the base with the uninfected branches on the same growth whorl, infected branches appear distinctly swollen. Large numbers of smaller branches or witches' brooms are often formed eventually on the infected branches. Broomed branches usually have dense foliage, outlive the noninfected branches around them, and continue to enlarge. Dwarf mistletoe aerial shoots form at some time during the enlargement of infected branches and are quite easy to recognize. Infections on small trees may result in rapid mortality. Large trees decrease in growth and progressively decline as witches' brooms develop. The foliage at the top of the tree becomes short, sparse, and yellowish and the upper crown begins to exhibit dieback. After an indefinite period of declining growth the entire tree dies, either as a result of the dwarf mistletoe parasite or due to the inability of the tree to prevent the invasion of weak parasites due to its low vigor. Infected, broomed branches are frequently the last to die.

Disease Cycle The aerial shoots of the dwarf mistletoe plants bear the flowers but they produce male and female flowers on separate plants (Fig. 7.1). During the mid to late spring male flowers produce pollen that is usually carried to the female flowers by insects (Fig. 7.2). After pollination fruit development begins on the female plants and it takes approximately a year for the fruit to reach maturity. When mature the fruits are oval berries that vary in size and color depending on

Fig. 7.3 Female dwarf mistletoe plant bearing mature fruit. (Photo courtesy of USDA Forest Service.)

Fig. 7.4   Dwarf mistletoe seed attached to a needle. (Photo courtesy of USDA Forest Service.)

the species of dwarf mistletoe (Fig. 7.3). Each berry contains a single seed and in mid-summer the mature seeds are dispersed by "explosive discharge." Water pressure that has built up inside the berries during maturation propels the discharged seed at high velocity [approximately 85 feet/sec (27 m/sec)] to a maximum horizontal distance of about 50 feet (16 m). The seeds are coated with a mucilaginous substance known as viscin and are able to adhere to most objects they contact in flight. Many of the seeds contact needles of adjacent trees (Fig. 7.4) while others will contact branches or trunks, or land on the soil. The seeds on needles remain attached where landed until a rainy period. Their mucilaginous coating absorbs water, swells, and becomes slippery (Fig. 7.5). The seeds slide down the needle until they reach the twig and then adhere to the bark (Fig. 7.6). Although infection is most common on small branches and twigs, occasionally older branches may be infected. The seeds of dwarf mistletoes remain dormant and germinate in the spring, with the exception of those species on ponderosa pine, which germinate within a month after dispersal. The root from the germinating seed grows until it reaches a suitable location for penetration of the host, forms a support structure (holdfast), and grows into the host bark (Fig. 7.7). Once the root reaches the living tissue in the inner bark it begins to branch out into a complex root network inside the bark and outer wood known as the endophytic system. This root system continues to enlarge throughout the infected branch and eventually may reach the main stem. Aerial shoots are usually produced 3 to 5 years after infection. Male and female plants usually reach sexual maturity 4 to 6 years after infection.

Treatment   Dwarf mistletoes are most effectively controlled by removing severely infected trees and by eradicating the pathogen in light to moderately infected trees by pruning. The most critical area to begin a sanitation program in a park, recreation area, residential area, or backyard is in the overstory. Infected trees in the overstory will result in the spread of the dwarf mistletoe seeds to the smaller understory trees below. Overstory trees should be removed if more than half of their branches or most of their upper crowns are infected. Trees less

Fig. 7.5   Mistletoe seed coat swollen with water. Note wet foliage. (Photo courtesy of USDA Forest Service.)

Fig. 7.6 Swollen seed of dwarf mistletoe that has slid down the fir needle and is in contact with the twig. (Photo courtesy of USDA Forest Service.)

severely infected should be pruned of all infected branches. Pruning cuts should be made at the trunk. When heavy pruning is necessary on a portion of the tree, isolated branches should not be left because they often contain latent infections. It is also recommended that, where the aesthetic value of the infected tree would not be damaged severely, two or three whorls of branches above the highest infected branch also be cut as well as all lower branches. This practice will decrease the necessity of the arborist to make follow-up pruning cuts if latent infections are present. However, infected trees should be watched for appearance of new infections each year. Remove infections as they appear.

Although heavily infected trees usually cannot be completely rid of infections, pruning can often help restore the vigor and prolong the life of valuable shade trees particularly when combined with a program of watering and fertilization. When infection of understory trees is not a prime concern, heavily infected shade trees in the overstory need not be removed if they are valuable ornamental trees. Trunk infections, although apparent indicators of extensive infection, are usually not serious on trees over 5 inch (13 cm) diameter because the endophytic system of the dwarf mistletoes spreads slowly in the main stem of larger trees and seldom results in mortality. On much smaller trees [under 2 inch (5 cm) diameter], however, the chance of mortality from trunk infections is much higher. Trunk infections may appear around pruning cuts where infected branches were removed. In these cases the endophytic system extended into the main stem before the infected branch was cut, but these infections are also not serious in larger trees.

Leafy Mistletoes

Leafy or "true" mistletoes are widely known because of their use for decorations during the Christmas season. The exact history of how the mistletoe custom began is lost in folklore and legend but it is known that since ancient times the evergreen mistletoe plants were believed to have magical and medicinal value. Most people, however, are unaware that these plants can also, on occasion, cause considerable injury and sometimes death of shade trees.

Leafy mistletoes, like the dwarf mistletoes, are flowering plants that require a living host tree to grow. However, unlike the dwarf mistletoes the leafy mistletoes produce true leaves and are able to manufacture most of their own food via

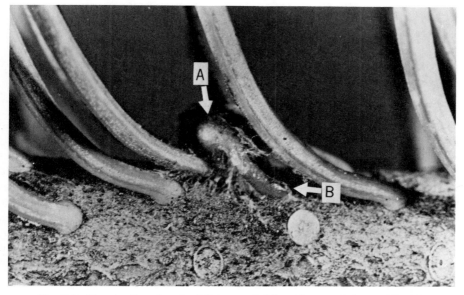

Fig. 7.7 Germinating dwarf mistletoe seed (A) with extended root (B). (Photo courtesy of USDA Forest Service.)

Fig. 7.8 Leafy mistletoe on oak during winter. (Photo courtesy of John M. Skelly, Virginia Polytechnic Institute and State University, Blacksburg, Virginia.)

avocado, citrus, cashew, cacao, coffee, mango, tea, and teak in Africa, Asia, and Central and South America.

Symptoms The evergreen mistletoe plants appear on the trunk and branches of susceptible species and are easily recognized after the host trees have shed their leaves in the fall (Fig. 7.8). The dark green leaves of the mistletoe plants are oval, rigid, and approximately 0.5 to 1 inch (12 to 25 mm) long (Fig. 7.9). Mistletoe plants often occur in clusters of stems that arise from one point (Fig. 7.10). Infection is commonly light with only a few clusters of plants on a tree; however, a tree may be so heavily infected with mistletoe that it may appear fully foliated in winter (Fig. 7.11). Swellings often occur on branches and on the trunk from mistletoe infections (Fig. 7.12).

Trees with light infection usually appear otherwise normal and healthy. Moderately to heavily infected trees, however, are often reduced in growth rate and photosynthesis. Water and minerals are all that is required from the host tree by the leafy mistletoe. The effect, therefore, of leafy mistletoe infection on the health of the tree, in most cases, is much less severe than that of dwarf mistletoes.

Hosts   Leafy mistletoes are found on a large number of hardwood trees and shrubs in the eastern, western, and southern United States. Their northern limit is roughly New Jersey to Oregon. Some commonly infected hardwoods include alder, ash, black locust, citrus, elm, hackberry, mahogany, maple, oak, pecan, persimmon, poplar, sycamore, walnut, willow, and yellow poplar. In Europe leafy mistletoes cause damage on a number of hardwoods including commercial fruit trees. Leafy mistletoes are also destructive on various tree crops such as

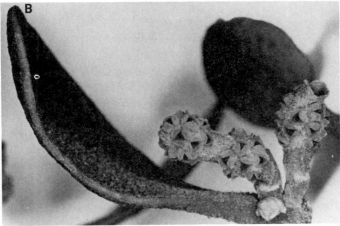

Fig. 7.9   (A) Shoots and mature fruit of leafy mistletoe (*Phoradendron serotinum*). (B) Close-up of leaf and female flower of *P. villosum*. (Photos courtesy of USDA Forest Service.)

Fig. 7.10 Extensive branching of leafy mistletoe plant from its origin on an oak branch. (Photo courtesy of John M. Skelly, Virginia Polytechnic Institute and State University, Blacksburg, Virginia.)

Fig. 7.11 Heavy infections of leafy mistletoe (A) on hackberry and (B) on California black oak. Photos taken in winter, foliage on tree due entirely to leafy mistletoes. (Photos courtesy of USDA Forest Service.)

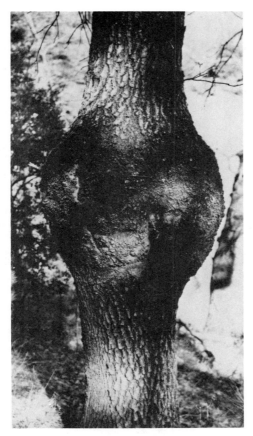

Fig. 7.12 Trunk swelling of California black oak from leafy mistletoe infection. (Photo courtesy of USDA Forest Service.)

in a state of decline. These trees are often predisposed to attack by insects. They are also more susceptible to drought or any other adverse environmental stresses than healthy trees of the same species around them. Heavily infected trees are often infected and sometimes even killed by weak parasites due to their low vigor.

Disease Cycle   White to pink berries are produced in clumps on female mistletoe plants (Fig. 7.13). Each berry contains a single seed that is enclosed in the pulp of the berry. Birds feed on the berries and ingest the intact seeds, which pass unharmed through the digestive system of the bird. The seeds that are deposited with the bird excrement will stick to a branch or trunk of a tree because of a sticky gelatinous coating that acts like cement when it dries. The seed germinates and the emerging root forms a haustorium that penetrates the bark and grows into the living tissues beneath it (Fig. 7.14). The root system of

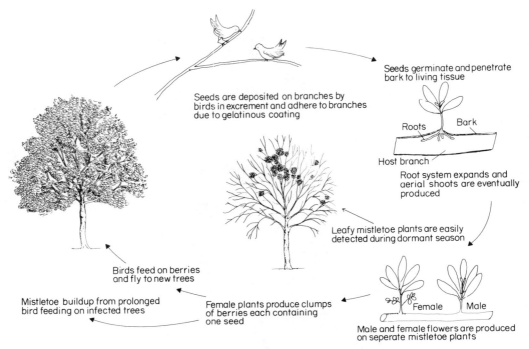

Seeds are deposited on branches by birds in excrement and adhere to branches due to gelatinous coating

Seeds germinate and penetrate bark to living tissue

Roots   Bark

Host branch

Root system expands and aerial shoots are eventually produced

Leafy mistletoe plants are easily detected during dormant season

Female   Male

Male and female flowers are produced on seperate mistletoe plants

Female plants produce clumps of berries each containing one seed

Birds feed on berries and fly to new trees

Mistletoe buildup from prolonged bird feeding on infected trees

Fig. 7.13 Disease cycle of leafy mistletoe on hardwoods caused by *Phoradendron* sp.

Fig. 7.14 Leafy mistletoe plant with the host wood removed to expose the haustorial (root) system in the tree. (Photo courtesy of John M. Skelly, Virginia Polytechnic and State University, Blacksburg, Virginia.)

Fig. 7.15 Bittersweet vine growing around a sugar maple (A) and wisteria vine (B) growing around a black cherry.

the pathogen expands up and down the living tissues of the infected branch. Aerial portions are produced when the root system has become extensive. Female flowers are pollinated by male flowers and seeds are produced.

Since birds tend to perch in the tops of taller trees it is the larger and older trees that are usually infected by the mistletoes. Consequently, smaller or young trees are rarely infected. Infections on an individual tree tend to build up as birds continually feed on mistletoe berries and deposit seeds on the branches and trunk.

Treatment Branches containing leafy mistletoe plants should be pruned at least 1 foot below the area of attachment of the mistletoe plants. If infections occur on large branches, the trunk, or wherever pruning is impractical, the

Fig. 7.15 *(continued)*

mistletoe shoots can be pruned or broken off. Removal of aerial parts does not eradicate the mistletoe but prevents it from obtaining needed food from photosynthesis. In about 2 or 3 years, however, the mistletoe will again produce new aerial shoots and their removal will again be necessary. Prolonged removal of aerial shoots has been achieved by removing a small patch of bark around the point of mistletoe attachment when the shoots are removed. Some success at inhibiting new shoot production after removal has also been achieved by wrapping tar paper or black plastic over the area where the shoots had been removed. Since most leafy mistletoe infections do little harm to the tree the treatment of these infections on shade trees is aimed primarily at improving the appearance of the tree. Consequently, this last method of treatment has met with limited acceptance with arborists. Some control of leafy mistletoes in forest trees and in commercial orchards has been achieved with herbicides applied during the dormant season but no information about their possible use for shade trees is currently available.

Other Parasitic Seed Plants

There are several additional parasitic plants that occasionally may parasitize shade trees. Some of the most common examples are broomrapes, dodder, false foxglove, Indian paintbrush, senna seymeria, and toothwort. The effect of these parasitic plants on trees is considered minor in most cases and control recommendations are not warranted.

NONPARASITIC SEED PLANTS

Introduction

Vines, which either shade out trees on which they are growing or strangle the trunk and branches, also act as pathogens to trees even though they are not parasites. These vines can be placed into two categories: (i) constrictors, and (ii) overgrowers. Constrictors injure trees by strangulation and deformation of the trunk and branches, while overgrowers shade out the host by growing completely over it.

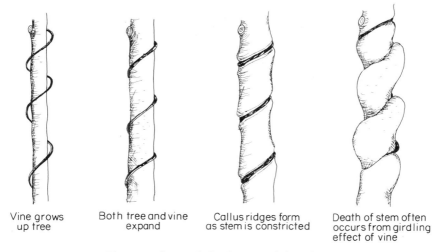

| Vine grows up tree | Both tree and vine expand | Callus ridges form as stem is constricted | Death of stem often occurs from girdling effect of vine |

Fig. 7.16 Strangulation by constricting vines.

Fig. 7.17 Trunk distortion from excessive callus production around a bittersweet vine on sugar maple (A) and a wisteria vine on a black cherry (B).

Constrictors

The most common examples of constricting vines that grow on shade trees are bittersweet, grape, and wisteria. Bittersweet and wisteria vines usually grow up the tree in snakelike fashion and may constrict a large amount of the trunk and branch area (Fig. 7.15). The affected tree produces callus around the vine as constriction increases (Fig. 7.16). Trees usually are not killed by the bittersweet or wisteria vines but the tree is disfigured (Fig. 7.17) and the movement of nutrients and water between roots and leaves is impaired (Fig. 7.18). Grape vines attach themselves to trees by means of tendrils. The constriction is, therefore, restricted to the periodic points of tendril attachment and is not as exten-

Fig. 7.17 *(continued)*

sive as that produced by bittersweet or wisteria vines. The constricted areas, which often appear gall-like from callus growth, are weak points and breakage may occur at these points. The control for problems on shade trees with either of these constrictors is complete removal of the vines or severing the vines at the soil where complete removal is unfeasible.

Overgrowers

Overgrowers are fast-growing vines that grow up the trunk and branches of trees and compete for sunlight (Fig. 7.19). These vines are particularly aggressive in open areas with abundant sunlight as is typical around most shade

Fig. 7.18   Bark killing on sugar maple caused by bittersweet vines.

trees. Some common examples of climbing vines that sometimes shade out trees are greenbriers, kudzu, Japanese honeysuckle, poison ivy, and Virginia creeper. These vines are usually most damaging to young trees, which are more easily overtopped by vines than older, larger trees. Control of overgrowers is similar to that of constrictors—removal of vines from shade trees, or severing of vines at the soil line. Some fast-growing vines, such as kudzu, have become a serious problem and can only be controlled with herbicides and controlled burning of infected sites.

Strangler Fig—A Constrictor and an Overgrower

The strangler fig, which is native to South Florida and the New World tropics, depends on other trees for support during its early development, but later

Fig. 7.19   Overgrowing vine (*Actinidia arguta*) on elm in the summer (A) and in the winter (B). (Photo courtesy of Shade Tree Laboratories, University of Massachusetts, Amherst.)

strangles and overgrows its host tree. Seeds of the strangler fig are deposited on trees in droppings of birds that have eaten the fruit of the strangler fig. The seeds germinate, and the young strangler fig lives as an epiphyte on its host tree. The cabbage palm is its most common host. As the young fig tree continues to develop it sends roots downward to the ground. Over the next several years the roots continue to enlarge and encircle the host tree, and at the same time the crown begins to crowd out the host tree's foliage (see Fig. 7.20). The host tree continues to decline from the combined effects of trunk strangulation and shading out of the foliage and eventually dies. Strangler figs growing as shade trees can often be seen with the dead and decaying stump of the host tree within their trunk.

SUGGESTED REFERENCES

Hawksworth, F. G., and D. Weins. (1972). Biology and classification of dwarf mistletoes (*Arceuthobium*). *U.S., Dep. Agric., Agric. Handb.* **401**, 1–234.

Lightle, P. C., and M. J. Weiss. (1974). Dwarf mistletoe of Ponderosa pine in the Southwest. *U.S., For. Serv., For. Pest Leafl.* **19**, 1–8.

Scharpf, R. F., and F. G. Hawksworth. (1974). Mistletoes on hardwoods in the United States. *U.S., For. Serv., For. Pest Leafl.* **147**, 1–7.

Fig. 7.20   A young strangler fig (right) grows around a cabbage palm (center).

Scharpf, R. F., and J. R. Parmeter, Jr. (1967). The biology and pathology of dwarf-mistletoe, *Arceuthobium campylopodium* f. *abietinum*, parasitizing true firs (*Abies* spp.) in California. *U.S., For. Serv., Tech. Bull.* **1362,** 1–42.
Skelly, J. M. (1972). American mistletoe. *Va. Polytech. Inst. State Univ. Co-op Ext. Serv.*, **MR-FTD-19,** 1–2.

8

Leaf Diseases

INTRODUCTION

A wide variety of disorders affect the leaves of shade trees. Perhaps no other part of the anatomy of a tree has been given more attention by those studying ornamental trees than the leaves. Leaf condition is often thought to be a primary indicator of the health of a tree. Although leaves sometimes accurately show tree health, in other cases a spectacular abnormality of the leaves may be of little importance to the health of the tree. The importance of a leaf disease on a shade tree depends upon the type of tree infected and its previous health history, as well as the type and severity of the disease. Since deciduous hardwoods and evergreens (broad- and narrow-leaved) differ largely in tolerance to leaf infections, these groups of trees will be discussed separately.

This chapter will deal primarily with leaf diseases caused by fungi. Leaf diseases caused by bacteria and viruses are covered in Chapters 3 and 5, respectively, on those disease agents. Many noninfectious stresses also cause leaf diseases; these are discussed in chapters on noninfectious diseases. Many insect conditions are commonly confused with leaf diseases. The causal insects often are inconspicuous or absent during inspection of the leaves. Common insect conditions will be discussed but require further study of appropriate entomological literature.

DECIDUOUS HARDWOODS

Leaf diseases of deciduous hardwoods, being conspicuous, are often brought to the attention of arborists by homeowners. Most hardwood leaf diseases, however, are among the least damaging diseases to shade trees. When the leaves of hardwoods are dropped each fall or winter, the effect of a foliar infection usually ends unless premature defoliation lowered food reserves. Yet control of a hardwood leaf disease is desirable. In addition, there are many other more serious diseases of other parts of the tree that also affect the foliage.

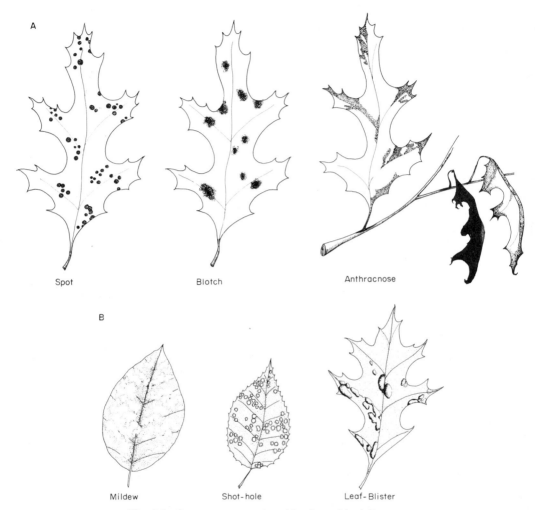

Fig. 8.1 Symptom categories of hardwood leaf diseases.

Because many species of fungi are capable of causing leaf diseases, and because there are many species of hardwoods, there are very many diseases. Hardwood leaf diseases, therefore, are grouped into the following six categories to simplify their identification (Fig. 8.1):

Leaf spot—dead area on the leaf that is well defined from healthy tissue
Leaf blotch—dead area on the leaf that often diffuses into the healthy tissues
Anthracnose—irregular dead areas on leaf margin, between and across and/or along veins, often moving onto the shoots and small twigs; sometimes whole leaves are engulfed

Powdery mildew—superficial growth of white to gray-white fungus material on
leaves and shoots

Leaf-blister—leaf spot or blotch that is swollen or raised, so that the area
appears blisterlike on the upper surface of the leaf

Shot-hole—loss of dead areas inside spots that results in a series of holes in the
leaf

The fungi that attack each tree species usually are different but often have
similar life cycles on hardwood leaves.

Life Cycles of Hardwood Leaf Fungi

Most fungal pathogens of hardwood leaves remain dormant in the dead
leaves on the ground or attached to the tree during the winter. Some pathogens
survive in the buds or on dead twigs and branches. In the spring during wet
weather the pathogens become active and discharge spores that are carried by
wind and rainsplash, and land on the young expanding leaves. The spores
quickly germinate and penetrate the leaves. By summer this infection produces
symptoms similar to one of the above six categories. During summer more
spores are produced in these infected areas, which can cause new leaf infections
during wet weather. These secondary spores are also spread by wind and rainsplash. Thus during wet growing seasons a considerable buildup of infections
on a tree can occur.

Treatment   Control measures for hardwood leaf diseases are usually recommended only when the health of the tree is poor or the aesthetic value of the tree
would be lost from the disease. The effect of a severe infection of a leaf disease
on an already stressed shade tree could tip the balance in favor of invading
microorganisms and could result in the loss of a valuable tree. In this case
control of leaf diseases is not only desirable but essential to survival of the tree.
Two major functions of trees planted around homes are the shade and the
beauty they provide to the homeowners; these also increase real estate values.
Although a leaf disease might not threaten the health of a shade tree it could
cause the tree to lose a great deal of its ability to cast shade and also change its
aesthetic effect from positive to negative. On shade trees, therefore, control of a
hardwood leaf disease can also be justified on aesthetic grounds. However,
since "beauty" is rather subjective, arborists will find that the importance given
hardwood leaf diseases varies considerably between clients.

Hardwood leaf diseases can be controlled in two principal ways: (1) protection
against infection with fungicides, and (2) use of resistant varieties. Fungicides
can be used effectively to protect the trees against infection from most leaf
pathogens by spraying at or before bud break and every 2 weeks following until
mid-June, with additional sprays if prolonged wet weather occurs in late spring.
The key to fungicidal control of leaf diseases is application of materials before
infection occurs and subsequent symptoms develop. Most homeowners do not

recognize leaf diseases until the symptoms have become extensive. They often request that sprays be applied to make the symptoms on the trees disappear. It is, of course, impossible to rid a tree of symptoms during the growing season since the affected leaf tissues are already dead. Homeowners should be told that this year it is only possible for further infection to be prevented by spraying, and that control efforts must be started early next year to prevent a recurrence of the disease. Resistant varieties are available for many of the tree species that are attacked by particularly troublesome leaf diseases. When new trees are to be planted, the arborist should choose or recommend these resistant varieties to avoid the needless future expense of spraying for clients. Use of resistant varieties is most warranted for street and park trees because they are always highly visible and financial resources for their care are minimal.

It has often been recommended that infected leaves be gathered and burned in the fall to remove any potential sources of inoculum for the tree. However, many leaf pathogens overwinter on the tree: in the buds, on the bark, or on dead twigs. Besides, the spores that infect leaves may also travel great distances in the wind, so that the removal of local sources of inoculum, such as from around the base of the tree, may have no effect on reducing disease severity. Moreover, dry leaves break up during raking and many pieces are left in the grass. Therefore, removal of fallen leaves gives little or no control of most leaf diseases.

The decision whether or not to use fungicide sprays to control leaf diseases of both hardwoods and evergreens is often difficult for both the arborist and the homeowner. A list of some of the pros and cons of chemical control of leaf diseases of shade trees is presented in Table 8.1.

SOME COMMON LEAF DISEASES OF DECIDUOUS HARDWOODS

Anthracnose

Hosts   Anthracnose is a common leaf disease that is found on a large number of tree species but is most severe on ash, maple, oak, and sycamore. This disease is common in most locations where these trees are grown but seems to be more severe near the northern edge of the normal range, northeast, and north central United States.

Symptoms   Anthracnose can be considered a transition disease between a leaf and a stem disease because it can often involve the leaves, shoots, buds, and twigs. In the spring, infected buds do not break because cankers have killed the bud tissues. During shoot elongation, infection results in rapid wilting and death of the expanding shoots and immature leaves. After shoot elongation, older and more mature leaf tissue often is killed progressively along the midrib or a vein, a pattern that is unique to anthracnose (Fig. 8.2). If the infection occurs early in the season and progresses rapidly leaf abscission may occur. Severe leaf

TABLE 8.1

Should One Spray a Tree To Try To Control Leaf Diseases?[a]

Reasons in favor of spraying

1. The tree is in a conspicuous place and the client therefore is willing to pay for the job just on the chance that the season might be wet and the tree might look bad from leafspots.

2. The tree has been weakened, for example by transplanting or by defoliation, in previous seasons, so it is improtant to insure against weakening again this season.

3. The tree is a species or variety that is usually hard hit by leaf spot diseases (for example, Paul's scarlet hawthorn), and so it is likely to be defoliated in case the season turns out to be a wet one.

4. The owner wishes to harvest fruit from the tree (in this case, a more elaborate spray program may be needed).

5. The tree is an evergreen and, being near a source of leaf spot inoculum, is in danger of defoliation (a single defoliation is fatal to evergreens, whereas it would only weaken a broad-leaved deciduous tree).

6. The tree is to be sold (for example, by a nursery) and only good-looking merchandise will sell.

Reasons against spraying

1. One is unable to tell in advance whether it will be a dry season, so that sprays applied at bud break might be wasted money.

2. The tree is large, healthy, and vigorous, and leaf spots of such a deciduous tree are not likely to matter to its overall health.

3. The client has only enough money for the most essential arboricultural work, and does not wish to spend money on the ultimate in tree beauty.

4. Once sprays are started, if the season continues wet longer and longer, the only way to get the value of the sprays is to apply additional ones, which would increase costs.

5. Some diseases (not others) can be at least partially controlled by raking up and destroying or composting last year's infected leaves. (However, others will blow in from a distance, so that cleaning one yard may make no difference.)

6. The arborist may have been contacted when the leaves have already begun to enlarge or have finished enlarging during wet weather, so that infections already would have taken place for that year.

7. Since leaf spot diseases are seldom fatal to deciduous trees, unless severe for many seasons in a row, there may be insufficient justification for putting more pesticide into the environment.

---

[a] Francis W. Holmes, Shade Tree Laboratories, University of Massachusetts, Amherst, Massachusetts.

abscission can result in defoliation (Fig. 8.3). The trees usually refoliate and assume normal appearance by midsummer. Cankers may occur on twigs and branches, where infections have killed buds or shoots. Cankers develop primarily in late winter and early spring. These cankers are small sunken dead areas that usually occur where an infected shoot joins a twig or an infected twig joins an older branch. Occasionally they girdle the branch. The excessive and angular branching of American sycamore and some white oak species are due primarily to chronic killing of twigs by the anthracnose disease (Fig. 8.4). These tree species serve as examples of how disease can alter normal tree form.

Fig. 8.2 Anthracnose on (A) ash; (B) elm; and (C) willow. (Photos courtesy of George N. Agrios, University of Massachusetts, Amherst.)

Fig. 8.3 American sycamore trees defoliated from anthracnose disease. Note tufts of foliage at top of both trees (A and B). (Photos courtesy of Shade Tree Laboratories, University of Massachusetts, Amherst.)

Fig. 8.4 Excessive and angular branching on American sycamore due to chronic twig killing from anthracnose. (Photo courtesy of Shade Tree Laboratories, University of Massachusetts, Amherst.)

However, there seem to be no chronic effects of anthracnose on the most commonly infected trees and many of these trees are extremely large and long-lived.

Disease Cycle   Anthracnose diseases of trees are caused by several species of fungi in the genus *Gnomonia*, but sometimes the fungi are also classified as *Gleosporium* spp. (Fig. 8.5). The fungus overwinters in infected buds and in fruiting bodies on dead leaves and twigs on the ground and on dead twigs and canker margins in the tree. Growth of the fungus usually begins in late winter or early spring during warm weather. At this time infected buds are killed, and cankers expand and sometimes girdle twigs and branches. During warm and wet weather in the spring, spores discharged from the fruiting bodies are blown or splashed into contact with emerging shoots. The infected shoots are killed and the fungus may grow into the attached twig causing a canker. These symptoms are sometimes confused with frost injury. Spores from fruiting bodies (Fig. 8.6) produced on the dead shoot can infect nearby expanding leaves. Spores germinate on leaf tissue and cause extensive necrosis and often defoliation

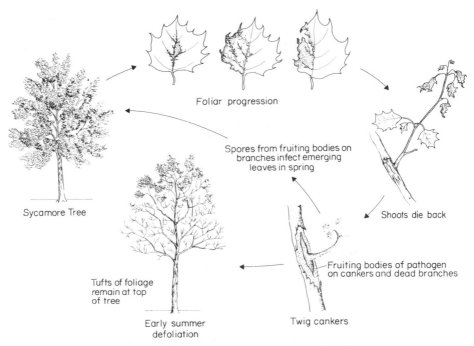

Foliar progression

Spores from fruiting bodies on
branches infect emerging
leaves in spring

Sycamore Tree

Shoots die back

Fruiting bodies of pathogen
on cankers and dead branches

Tufts of foliage
remain at top
of tree

Early summer
defoliation

Twig cankers

Fig. 8.5   Disease cycle of sycamore anthracnose caused by *Gnomonia veneta*.

during late spring and early summer. Fruiting bodies are also produced on necrotic leaf tissue. Disease development is minor in the summer and into the fall when infected leaves and some dead shoots and twigs are dropped.

Treatment   Anthracnose diseases are difficult to control and they are seldom severe enough to warrant control measures. Severe defoliation for 2 or 3 successive years, however, can greatly decrease the health of trees and make them more susceptible to numerous environmental stresses and to secondary pathogens. Perennial infections of anthracnose may also decrease the growth and attractiveness of a valuable ornamental tree. In these cases symptoms can be reduced significantly with a yearly program of fungicide applications. Three applications are needed each spring: the first when the buds are about to break, the second 2 weeks after the first, and the third when the leaves are about one half mature size. Adding a spray in autumn, after leaf fall, will greatly increase control. It has also been suggested that removal of dead and cankered twigs and branches from the tree and removal of fallen leaves will reduce infection the following year. However, since there is evidence that the fungus on fallen leaves does not contribute to new infections, their removal will not be much help. Application of balanced fertilizers and watering during dry periods will help the tree to recover strength after severe infections.

Anthracnose diseases can be prevented in many cases by the avoidance of highly susceptible species such as American sycamore and white oak. London plane, a species resistant to anthracnose, is planted extensively as a substitute for American sycamore. The two species are so similar that most people have difficulty distinguishing between them, but the exposed patches on sycamore trunks are much whiter. Many anthracnose-resistant oak species, particularly those in the red oak group, can be substituted for highly susceptible white oaks in areas where severe anthracnose is a perennial problem. In areas where oak wilt is common, however, red and black oaks are more severely attacked by oak wilt than are white oaks (see Chapter 12, Wilt Diseases).

Dogwood Anthracnose

Decline of flowering dogwood in both forest and landscape sites has become a serious problem in the northeast United States. Affected trees exhibit

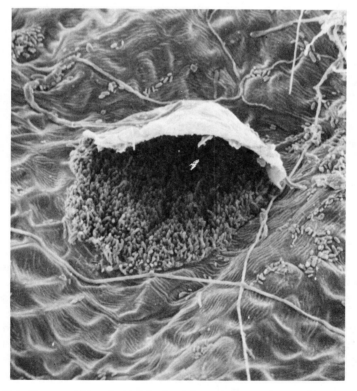

Fig. 8.6 Spores within a fruiting body of the anthracnose fungus *Gleosporium apocryptum*. (SEM photo courtesy of Merton F. Brown and H. G. Brotzman, University of Missouri, Columbia.)

pronounced lower branch dieback, which usually leads over several years to progressive whole crown dieback and death of the tree. Initially called "dogwood lower branch dieback," this disease has been found to be consistently associated with an anthracnose fungus, *Discula* sp., and is now called dogwood anthracnose (Daughtrey, Hibben, and Hudler 1986). This disease attacks leaves, shoots, and branches. In contrast to other anthracnose diseases of trees, dogwood anthracnose can be a serious health threat to susceptible trees.

Hosts   Flowering dogwoods (all cultivars) are commonly attacked. Kousa dogwood, a widely used Asian import, is low in susceptibility.

Symptoms   Twig and branch death occur on the lowest branches, which results in the "lower branch dieback" syndrome. The dieback becomes progressively more extensive over several seasons and may eventually involve the entire crown. On the foliage, small round spots and irregular blotches, both with purple borders, occur. The dead centers of the spots and blotches may eventually drop out, leaving a "shot hole" symptom on the affected leaves. Water sprouts may form on the trunk and main branches in later stages of the disease. However, these are often infected and die back. Small areas of the bark are often killed at the base of the sprouts from cankers that often develop at these points. Chronically infected trees become severely weakened and are often attacked by dogwood borers. The combination of low vigor from several years of the anthracnose disease and the attack by secondary borers is usually fatal to landscape dogwoods.

Control   Maintaining dogwood trees in a vigorous condition has been the best management strategy for the control of dogwood anthracnose. Avoid moisture stress by regular watering. Encourage growth through regular fertilization. Prune dead twigs and all water sprouts (sucker growth) each spring and fall. Avoid injuries to trees from lawn care equipment. Broad-spectrum fungicides have sometimes been used to enhance these cultural controls, but their effectiveness still has not been demonstrated.

Scab

Hosts   Scab is an important leaf and fruit spot on commercial apples and pears grown for fruit production. It can also be a problem on ornamental apples, crabapples, pears, hawthorns, and other closely related members of the apple family that are used for shade trees. A closely related disease also occurs on willow.

Symptoms   Circular lesions appear on the expanding leaves in the spring. The lesions are dark green at first and progressively turn green-black and then black (Fig. 8.7). As leaves mature the necrotic areas cause curling and dwarfing in severe infections. Lesions on a leaf may either remain separate or may merge into one large lesion. Developing fruit may also develop lesions, which become thickened and rough or "scabby." Mature fruits may be distorted or cracked.

Fig. 8.7 Scab on an apple leaf. (Photo courtesy of Shade Tree Laboratories, University of Massachusetts, Amherst.)

Disease Cycle Scab diseases are caused by species of fungi in the genus *Venturia* (Fig. 8.8). *Venturia inaequalis* causes apple scab, *V. pyrina* causes pear scab, and *V. saliciperda* causes willow leaf blight. The pathogen overwinters as immature fruiting bodies on dead leaves on the ground. During wet weather, at the approximate time of bud break, the fruiting bodies become fully mature and forcibly discharge spores into the air. These spores contact and infect the emerging leaves. The fungus grows within the leaf tissue causing a lesion. Secondary spores are produced usually within 2 weeks of the initial infection. Spores of this repeating stage cause new infections on leaves and developing fruit throughout the growing season. The amount of spore production and infection rate are favored by wet, cool conditions. Infected leaves and fruit drop to the ground in the fall and form immature fruiting bodies that will produce the spores for next spring's infections.

Treatment Scab diseases of shade and ornamental trees rarely merit control measures although commercial fruit trees are sprayed with fungicides many

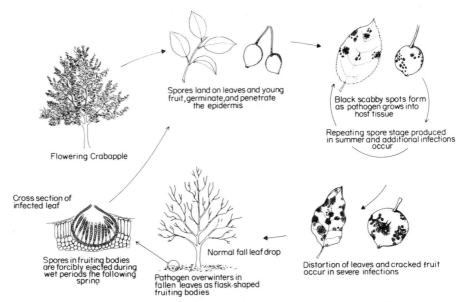

Spores land on leaves and young
fruit, germinate, and penetrate
the epidermis

Black scabby spots form
as pathogen grows into
host tissue

Repeating spore stage produced
in summer and additional infections
occur

Flowering Crabapple

Cross section of
infected leaf

Spores in fruiting bodies
are forcibly ejected during
wet periods the following
spring

Normal fall leaf drop

Pathogen overwinters in
fallen leaves as flask-shaped
fruiting bodies

Distortion of leaves and cracked fruit
occur in severe infections

Fig. 8.8 Disease cycle of apple scab caused by *Venturia inaequalis.*

times each year for protection against losses from these diseases. Since fruit production is either nonexistent or of minor consequence on ornamental trees, the major concern of the homeowner is the leaf spot. Leaf spots usually do little harm to health and appearance of the tree. Control of leaf symptoms, however, can be achieved with 3 to 4 applications of fungicides, at 10-day intervals, beginning just before bud break. Additional sprays may also be necessary during wet springs or when heavy rains occur between spray applications.

Powdery Mildews

Hosts Powdery mildews have a very large host range that includes many woody and herbaceous plants. Some of the trees commonly infected with powdery mildews are apple, ash, beech, birch, boxwood, buckeye, catalpa, cherry, crabapple, dogwood, elm, euonymus, hackberry, hawthorn, hickory, holly, honey locust, horse chestnut, lilac, linden, maple, oak, pear, planetree, plum, poplar, rose, sycamore, tuliptree, walnut, and willow.

Symptoms During mid to late summer foliage develops patches of gray-white material that resembles powder or dust on top of the leaves (Fig. 8.9). These patches enlarge and often may cover the entire leaf (Fig. 8.10). In some cases a white "powder" may be dislodged from the leaf during handling. The patches are most abundant and develop most rapidly on leaves in the lower branches and in heavily shaded branches. Heavily infected foliage may be discolored, distorted, or stunted, but light to moderately infected leaves may appear otherwise normal. During the fall small brown to black bodies may appear within the gray-white areas. These specks often look like ground black pepper in the white powder on the leaf. On many host species, however, powdery mildews do not produce these black specks until after the leaves fall. Severe perennial infection of powdery mildew can cause twig dieback and significant growth retardation, but most trees are not harmed by this disease.

Disease Cycle   Powdery mildews are caused by many species of fungi in six genera, *Erysiphe, Microsphaera, Phyllactinia, Podosphaera, Sphaerotheca,* and *Uncinula.* The pathogens overwinter in the black specks, which are fruiting bodies, or as vegetative mycelium in buds and twigs. In the spring, spores are released from the fruiting bodies and are blown or splashed onto the foliage of susceptible trees. The pathogen penetrates the leaf epidermis and inserts into the spongy mesophyll or the palisade cell a specialized absorbing organ, called a haustorium. The fungus does not grow further into the leaf but obtains its

Fig. 8.9   Light infection of powdery mildew. (Photo courtesy of Shade Tree Laboratories, University of Massachusetts, Amherst.)

nutrition from the leaf through the haustoria. This structure does not kill the cell and the fungus can only live as long as the penetrated cell remains alive. As the fungus grows extensively on the leaf surface the leaf appears gray-white. Soon after the fungus extends over the leaf surface it begins to produce long chains of spores. These secondary spores, sometimes called "summer spores," create new infections during the late spring and throughout the summer. Best growth of the pathogen occurs during warm, damp, or wet weather. When cool weather occurs in early fall, production of summer spores ceases and the fruiting bodies of the overwintering stage begin to form on top of the leaf (Fig. 8.11). The fruiting bodies mature on the dead leaves during the winter.

Treatment   Powdery mildews on the foliage of deciduous trees rarely merit control measures outside the nursery, but this disease can cause considerable damage to flowers and is a major problem on roses. However, some susceptible tree species, such as lilac, can lose much of their ornamental value when severely infected. In the nursery and greenhouse powdery mildew is often controlled with fungicides. They are applied at 1- or 2-week intervals after symptoms are observed, depending on the severity of the problem. Similar recommendations apply to shade trees.

Powdery mildews can be minimized by providing adequate sunlight and air circulation around plants in the greenhouse, nursery, and around the home.

Fig. 8.10   Powdery mildew on (A) lilac and (B) catalpa.

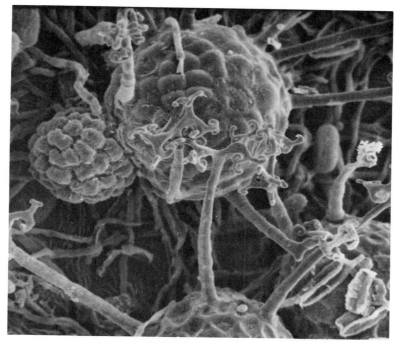

Fig. 8.11 Overwintering fruiting bodies of the powdery mildew fungus *Microsphaera alni*. (SEM photo courtesy of Merton F. Brown and H. G. Brotzman, University of Missouri, Columbia.)

Overcrowding, watering late in the day, and planting small trees and shrubs in the dense shade of larger trees all favor development of powdery mildew. Some tree species, such as lilac and many oaks, develop powdery mildew each year when leaves begin to senesce in late summer or early fall. Although the symptoms are highly visible, the disease causes little harm to trees, which will soon drop the infected leaves. In these typical and common cases on shade trees usually no attempt is needed to minimize or control this disease. Sometimes homeowners may request spray applications to improve the appearance of valuable trees or shrubs.

Oak Leaf Blister

Hosts   Most oak species are susceptible but pin oak and the white oak group are most susceptible. Closely related leaf blisters and "curls" also occur on a large number of other species such as birch, California buckeye, elm, maple, pear, poplar, and willow.

Symptoms   In the early summer local areas of tissue grow faster than the rest of the leaf and cause the illusion of blisters on the leaf (Fig. 8.12), although the

Fig. 8.12   Leaf blister on red oak. (Photo courtesy of George N. Agrios, University of Massachusetts, Amherst.)

opposite surface is curved inwards and there is no accumulation of fluid. The blisters are often lighter green than the normal leaf tissue. Severe infections often result in midsummer defoliation.

Disease Cycle   Oak leaf blister is caused by the fungus *Taphrina caerulescens*. Other closely related leaf blister diseases are caused by other species in the genus *Taphrina* and have a disease cycle similar to oak leaf blister. The pathogen overwinters as spores attached to buds of susceptible trees. Spore germination occurs in the spring at the time of bud break, and the fungus enters the newly emerging leaves through the stomates. The fungus grows within the leaf and causes the tissues in the area of infection to grow much faster than the rest of the leaf. The leaf tissue soon becomes distorted and forms a bump or blister on the leaf. The fungus produces spores within the leaf, which eventually become exposed when the leaf epidermis ruptures. The spores are discharged in the fall and may be blown to the newly formed buds.

Treatment   Leaf blister diseases, including oak leaf blister, are rarely severe enough to require control measures. Since severe infections can cause defoliation of shade trees, protection of foliage with fungicides is sometimes requested by the homeowner. One application of fungicide before bud break will control leaf blister but applications after bud break are usually not effective.

Tar Spot

Hosts   Numerous species of maples and willows.

Symptoms   Small yellow spots occur on the upper leaf surface during mid to late July. The spots increase in size and turn black (Fig. 8.13). The spots usually appear as single dark areas but may occur as a local concentration of many small spots as on mountain maple. The black spots are thicker than the normal leaf tissue and sometimes become ridged later in the season. Infected leaves are held on the tree until normal leaf drop in the fall.

Disease Cycle   Tar spot is caused by species of fungi in the genus *Rhytisma*— on maple *R. acerinum*, on mountain maple *R. punctatum*, and on willow *R. salicinum* (Fig. 8.14). The fungi overwinter as immature fruiting bodies in the spots on fallen leaves. The fruiting bodies mature during the spring. Spores are released and contact the expanding foliage of susceptible trees. The fungus grows within the leaf tissue and produces a black gummy substance that causes the fungus and host tissue to stick together. The combination of host tissue and

Fig. 8.13   Tar spot on Japanese maple.

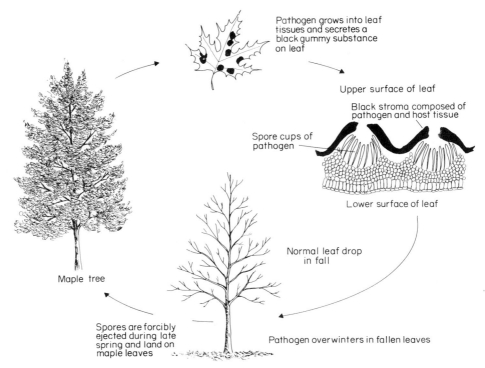

Fig. 8.14   Disease cycle of tar spot of maple caused by *Rhytisma acerinum.*

fungus is referred to as a *stroma*. The upper epidermis of the leaf is pushed outward as the stroma continues to develop during the summer. Fruiting bodies of the fungus are formed within the stroma that will become the primary inoculum for the next season (Fig. 8.15). In the fall infected leaves containing the immature fruiting bodies are cast.

Treatment   Tar leaf spot has minor impact on the health of the tree but has a quite striking appearance. This disease can be controlled by applications of fungicides at 2-week intervals during leaf emergence, but control treatments are rarely applied.

Hawthorn Leaf Spot or Blight

Hosts   Several varieties of hawthorn are susceptible as well as apple, ash, cotoneaster, mountain ash, pear, and quince. English hawthorn and Paul's scarlet hawthorn are highly susceptible while Cockspur thorn and Washington thorn appear to be resistant.

Symptoms   Small red-brown spots appear on the leaf during leaf expansion in early spring. The spots enlarge and join together as the season progresses and the leaves become covered with spots. In mid to late summer severely infected leaves drop and the tree may be completely defoliated (Fig. 8.16).

Disease Cycle Hawthorn leaf blight is caused by the fungus *Entomosporium maculatum*. The pathogen overwinters on the tree, as fruiting bodies on twigs and bark and on fallen leaves. During wet weather in early spring spores are discharged that cause infections on the expanding foliage. The fungus grows in the leaf tissue and kills spots. Continued growth of the fungus causes the leaves to fall.

Treatment Severe infection and defoliation can be prevented by avoiding highly susceptible varieties of hawthorn. Most varieties receive mild infection during wet seasons with little effect on the tree's health. Severe infections for several consecutive seasons, however, can weaken a tree and control measures are warranted in this case. Control sufficient to prevent defoliation can be achieved with three applications of fungicides at 10-day intervals beginning just after the leaves have unfolded.

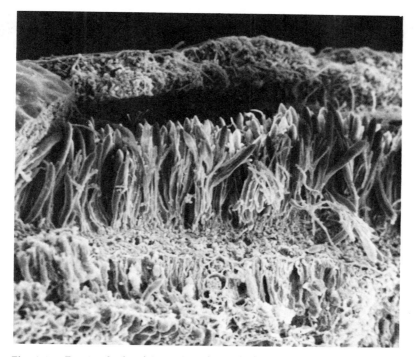

Fig. 8.15 Fruiting body of *R. acerinum* beneath the stroma caused by the tar spot infection. (Photo courtesy of Merton F. Brown and H. G. Brotzman, University of Missouri, Columbia.)

Fig. 8.16 Hawthorn leaf blight. (A) Defoliation from severe infection. (B) Close-up. Note a few symptomatic leaves still remain on the tree.

Other Leaf Diseases of Deciduous Trees

Only a small fraction of the large number of leaf diseases of deciduous trees that exist have been discussed. It is reasonable to expect to encounter at least one infectious leaf disease for each species of tree and shrub, and often more than one. There are, however, many similarities among the symptoms, among the life cycles, and among the treatments required for most of these diseases. Anyone working extensively with woody plants will often encounter unfamiliar leaf diseases. Much of this information about the diseases that have been discussed can, therefore, be applied to unfamiliar diseases, and similar control measures can be tried.

BROAD- AND NARROW-LEAVED EVERGREENS

In contrast to leaf diseases of hardwoods, diseases that affect the leaves of conifers and broad-leaved evergreens often are dangerous to plant health. They can even kill trees and shrubs. Evergreens have adapted to holding their leaves the entire year and individual leaves often persist for several years. The infection and subsequent loss of these leaves results in a much greater physiological shock to the plant when compared to the loss of leaves in deciduous trees. Evergreens with severe leaf infections are also more susceptible to attacks by other pathogenic microorganisms, insects, and environmental stresses. There is also a more prolonged aesthetic loss when an ornamental evergreen exhibits symptoms of leaf infection or suffers the loss of its foliage as compared to deciduous trees that normally become void of foliage in the fall. Deciduous conifers, such as bald cypress and larch, can be included with evergreen conifers because they have many of the same type of needle diseases.

Evergreen leaf diseases like hardwood leaf diseases are caused by numerous species of fungi. Therefore, there are also many individual leaf diseases of evergreens. Most of these diseases, however, can be placed into one of the following categories:

Leaf spot—Broad-leaved evergreens exhibit spot symptoms that are quite similar to those found on deciduous trees.

Needle spot, needle cast, needle blight—These symptoms are all closely related to the total disease syndrome of most leaf diseases of conifers. Individual infections occur at random areas along the needle and result in the formation of a dead area or spot. When the infection from one or more spots spreads over the needle, the entire needle is progressively killed and the tree responds by shedding the infected needle. The result is a pile of dead needles on the ground around the base of the trunk and infected and dying needles on the tree. This condition is usually called a "needle cast," and when it appears quite severe it may be called a "needle blight," but these two terms are often used synonymously.

Needle rust—Swollen white or orange blisters occur on the needles. Occasionally rust-infected needles are cast from the tree but most infected needles remain attached. Needle rusts will be discussed in detail in Chapter 11, Rust Diseases.

Life Cycle of Evergreen Leaf Fungi

Most pathogenic fungi on evergreens overwinter in a dormant state in infected leaves both on the tree and on the ground. Some pathogens, however, are able to overwinter in buds or in dead twigs or branches. In the spring or summer the pathogens mature and discharge spores during wet weather. Spores are carried by wind and/or rainsplash to the current year's foliage. Infection may also be spread by infected nursery stock and on clothes and tools. Germination and penetration occur when foliage is wet. Symptoms similar to one of the common categories of evergreen leaf diseases occur eventually. A secondary (repeating) spore stage during the growing season may or may not be present.

Treatment   Most control efforts are aimed at preventing infection of the current year's foliage. Control recommendations can vary with each disease since foliage infection can occur, in some cases, anytime from around budbreak to late fall. Several chemical spray applications may be needed for effective control and additional ones may also be needed during wet seasons.

Some degree of control of many evergreen leaf diseases can often be achieved by steps taken to minimize the amount of time that foliage remains wet after rains. Often shade intolerant evergreens are placed by homeowners in excessive shade. The foliage becomes exposed to high humidity conditions and consequently becomes highly susceptible to leaf infections. Prune and plant evergreen shrubs and trees to allow air movement between plants and to allow enough sunlight to aid in drying foliage after wet periods. Homeowners may also contribute inadvertently to the incidence and severity of leaf diseases of both hardwoods and conifers during watering if they spray the foliage. Water should only be applied to the soil under the branches of trees and shrubs.

SOME COMMON EVERGREEN LEAF DISEASES

Azalea Leaf Gall

Hosts   Azaleas, blueberries, cranberries, and rhododendrons are all susceptible but this disease is most commonly found on azaleas.

Symptoms   Swollen and thickened areas occur on the newly developing leaves during the spring. These gall-like areas are pale green to white and glossy. The entire leaf or only a part of it may be affected. Flower buds may also become affected and the entire flower develops into an enlarged waxy gray-white gall (Fig. 8.17). The leaf and flower galls turn brown with age and later dry out. Infected leaves usually abscise during the summer. Severe infection can result in partial defoliation of a shrub and the loss of most of the flowers.

Disease Cycle   Azalea leaf gall is caused by the fungus *Exobasidium vaccinii* (Fig. 8.18). This pathogen overwinters as spores on bark or budscales. During bud swell and bud break the spores germinate and infect the newly emerging leaves or flowers. The fungus grows in the host tissue and stimulates it to produce galls. In the summer the fungus produces spores on the surface of the galls that can be spread to the newly formed buds.

Treatment   Control measures usually are not needed outside the nursery unless severe cases develop. This disease can be prevented by applications of fungicides starting at bud break and continuing at 10- to 14-day intervals until leaves mature. Applications following symptom appearance will not be effective since the pathogen is already within the host tissues. Removal by hand of galled plant parts is the only recommended treatment once symptoms occur. The disease is favored by prolonged wet foliage that is usually associated with restricted air movement and heavy shade. Disease severity, therefore, may also be

Fig. 8.17 Azalea leaf gall on flower buds of rhododendron. (Photo courtesy of George N. Agrios, University of Massachusetts, Amherst.)

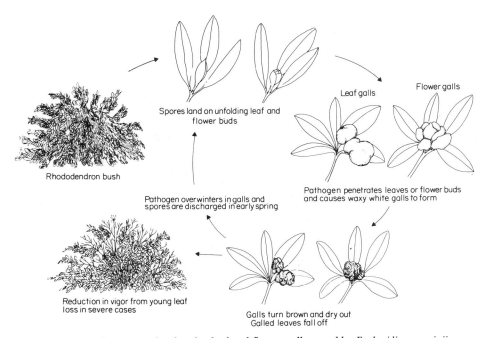

Spores land on unfolding leaf and flower buds

Leaf galls   Flower galls

Rhododendron bush

Pathogen penetrates leaves or flower buds and causes waxy white galls to form

Pathogen overwinters in galls and spores are discharged in early spring

Reduction in vigor from young leaf loss in severe cases

Galls turn brown and dry out
Galled leaves fall off

Fig. 8.18 Disease cycle of azalea leaf and flower gall caused by *Exobasidium vaccinii*.

decreased by limiting shade and facilitating air movement with increased spacing between plantings.

Lophodermium Needle Cast

Hosts A large number of pine species are susceptible including Austrian, digger, eastern white, jack, jeffrey, knobcone, limber, lodgepole, longleaf, monterey, pitch, ponderosa, red, Scots, shortleaf, slash, spruce, sugar, and whitebark pines. Most seriously infected, however, are red and Scots pine. Short-needled varieties of Scots pine such as Spanish and French Green are highly susceptible while long-needled varieties such as Scottish Highland and Austrian Hills have greater resistance.

Symptoms In the early spring small brown spots, usually with yellow halos around them, occur on last year's needles. These spots enlarge and entire needles begin to turn yellow then brown in late spring (Fig. 8.19). Disease severity is usually worse on the lower half of the tree but the entire tree may be uniformly

Fig. 8.19 Lophodermium needle cast on Scots pine. (Photo courtesy of Darroll D. Skilling, USDA, Forest Service, St. Paul, Minnesota.)

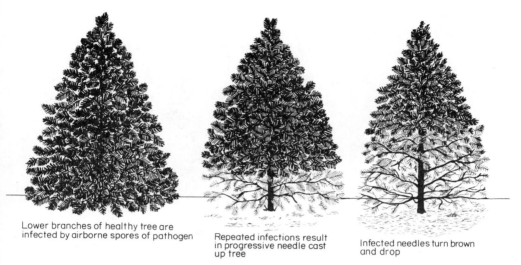

Lower branches of healthy tree are infected by airborne spores of pathogen

Repeated infections result in progressive needle cast up tree

Infected needles turn brown and drop

Fig. 8.20 Disease progression of Lophodermium needle cast.

infected. Brown needles begin to drop in early summer and continue to do so throughout the summer (Fig. 8.20). Most dead needles are cast but some remain attached to the tree. In late summer small black football-shaped structures, each with a lengthwise slit in the middle, occur on the dead needles. In severe infections most of the previous year's foliage is cast and only the current year's needles remain in the fall.

Disease Cycle Lophodermium needle cast is caused by the fungus *Lophodermium pinastri*. The fungus overwinters on the tree in infected current year's needles. During the spring the fungus resumes growth in the infected needles and causes death of leaf tissues. Infection usually results in the death of needles by late spring. During the early summer fruiting bodies develop on dead needles (Fig. 8.21). Spore discharge begins in late summer and continues into mid-fall, especially during wet weather. The spores are forcibly discharged into the air and contact the current year's foliage. Infection occurs and the pathogen remains as vegetative mycelium in the leaf's tissue during the winter.

Treatment Lophodermium needle cast is a serious disease of pines and can cause severe losses in both ornamental and forest tree nurseries. Larger trees usually survive severe infection, but often lose many of their lower branches and, hence, are greatly reduced in value as shade trees. Small trees, however, are sometimes killed or are severely damaged, unattractive, and unsalable for several years.

The most effective control of this needle cast disease can be achieved by fungicide applications during peak periods of spore release (Fig. 8.22). Applications should begin about August 1 and continue at 2-week intervals until mid-September. Additional sprays may be required during warm, wet fall weather.

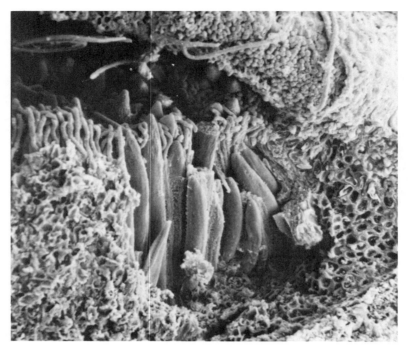

Fig. 8.21 Fruiting body of *Lophodermium pinastri* cut away to expose spore sacs (×750). (Photo courtesy of Merton F. Brown and H. G. Brotzman, University of Missouri, Columbia.)

Fig. 8.22 Spraying Scots pine with a mistblower to control needle cast diseases. (Photo courtesy of Darroll D. Skilling, USDA, Forest Service, St. Paul, Minnesota.)

In the nursery the following cultural precautions will also minimize disease severity: (1) examine all incoming plants to ensure healthy nursery stock, (2) avoid highly susceptible varieties of Scots pine nursery stock, (3) do overhead watering only in the morning to avoid prolonged wet foliage, (4) avoid Scots or red pine for windbreaks because they may be sources of inoculum, and (5) avoid monocultures of one species or variety that could all become victims of a disease epidemic of Lophodermium needle cast or any other infectious disease. Some of the same cultural practices will also be beneficial around the home, such as inspection of new plantings, avoiding highly susceptible varieties, avoiding wet foliage for prolonged periods, and avoiding monocultures. In addition, these sound cultural preventative measures will also minimize disease severity in other fungus leaf diseases that infect evergreens.

Juniper Twig Blight

Hosts   Susceptible species include arborvitae, cypress. Douglas fir, eastern red cedar, fir, hemlock, juniper, larch, redwood, white cedar, and yew. Eastern red cedar and other junipers are the most commonly infected. A large number of resistant varieties of juniper have been reported by Schoeneweiss (1969), a selection of which appear in Table 8.2.

Symptoms   During the spring tips of twigs and branches turn light green and then brown. Twigs and branches less than $\frac{1}{3}$ inch (less than 1 cm) diameter are most commonly affected. In the summer the dead foliage turns ash-gray and small black spots appear on the gray tissue. Close inspection reveals cankers at the junction between dead and live tissue. After several years, in severe cases, many of the lower branches are killed and the attractiveness of the tree or shrub is markedly decreased.

Disease Cycle   Juniper twig blight is sometimes called Phomopsis blight because it is caused by the fungus *Phomopsis juniperovora* (Fig. 8.23). This fungus overwinters in infected tissues as fruiting bodies and vegetative mycelium. During wet weather in the spring and in late summer or early fall spores ooze out of the fruiting bodies and are blown or splashed to healthy foliage. Infection also depends on wet foliage and, therefore, little infection occurs during dry periods in the summer. The pathogen grows into twigs and causes cankers that eventually girdle the stem. The tissue beyond the canker dies and the pathogen grows extensively in the dead tissue. Fruiting bodies form in the dead tissue during early summer; these can ooze spores in wet periods later in the season, and create new infections. The fungus can remain alive in dead tissue and produce fruiting bodies for at least 2 years.

Treatment   The following cultural practices will decrease disease severity: (1) infected branches should be pruned out during dry weather; (2) pruning and handling plants during wet weather should be avoided; (3) good ventilation between plants should be provided to promote rapid drying after rains; (4) planting susceptible species in heavy shade should be avoided. Control of

TABLE 8.2
Junipers Resistant to Twig Blight[a]

Common name	Species	Variety[b]
Chinese juniper	*Juniperus chinensis*	c.v. Femina
Chinese juniper	*Juniperus chinensis*	c.v. Iowa
Chinese juniper	*Juniperus chinensis*	c.v. Keteleeri
Chinese juniper	*Juniperus chinensis*	c.v. Pfitzeriana Aurea
Chinese juniper	*Juniperus chinensis*	var. *Sargentii*
Chinese juniper	*Juniperus chinensis*	c.v. Shoosmith
Common juniper	*J. communis*	c.v. Ashfordii
Common juniper	*J. communis*	c.v. Aureo-spica
Common juniper	*J. communis*	var. *depressa*
Common juniper	*J. communis*	c.v. Hulkjaerhus
Common juniper	*J. communis*	c.v. Prostrata Aurea
Common juniper	*J. communis*	c.v. Repanda
Common juniper	*J. communis*	var. *Saxatilis*
Common juniper	*J. communis*	c.v. Suecica
Creeping juniper	*J. horizontalis*	c.v. Depressa
Creeping juniper	*J. horizontalis*	c.v. Procumbens
Savin	*J. sabina*	c.v. Broadmoor
Savin	*J. sabina*	c.v. Knap Hill
Savin	*J. sabina*	c.v. Skandia
Western red cedar	*J. scopulorum*	c.v. Silver King
Western red cedar	*J. squamata*	c.v. Campbellii
Western red cedar	*J. squamata*	var. *Fargesii*
Western red cedar	*J. squamata*	c.v. Prostrata
Western red cedar	*J. squamata*	c.v. Pumila
Red cedar	*J. virginiana*	c.v. Tripartita

[a] Schoeneweiss (1969) and Peckhold *et al.* (1976)
[b] c.v. = cultivar; var. = botanical variety.

juniper twig blight can be achieved most successfully with fungicide applications and the use of resistant varieties. Applications should be made once every 2 weeks from the onset of shoot growth in the spring to early fall; however, applications are not needed during summer dry periods.

Brown Spot Needle Blight

Hosts   Many species of pine are susceptible including Austrian, digger, eastern white, knob cone, lodgepole, longleaf, pitch, pond, ponderosa, Scots, short-leaf, slash, spruce, and western white pine. This disease is most severe on longleaf and Scots pine. As with Lophodermium needle cast, short-needled varieties of Scots pine are the most susceptible.

Symptoms   During late summer small gray-green to gray-black spots appear on needles. The spots later turn brown in the center and are surrounded with a yellow halo. The spots increase in size until the needles are killed the same season. These symptoms are similar to those found in Lophodermium needle cast; however, in that disease the symptoms occur in the spring. Dead needles are cast during the fall. Small black fruiting bodies form on the dead needles. These fruiting bodies, at first, may appear similar to those produced by *L. pinastri,* but these fruiting bodies are not shaped like split footballs and only protrude slightly from the needle surface. The death of needles is usually restricted to the bottom half of the tree (Fig. 8.24). Several years of infection results in severe damage or death of seedlings and young trees, and the loss of aesthetic value on older trees.

Disease Cycle Brown spot is caused by the fungus *Scirrhia acicola.* The fungus overwinters as fruiting bodies in dead needles. In the spring the fruiting bodies begin to discharge spores during periods of wet weather. Spore production increases progressively until late summer. The spores contact current-year needles and cause infections that eventually result in the death of the needles. The fungus develops fruiting bodies in the dead needles that will produce spores for the following year's infections. A second type of spore has been found in the southern United States that is wind blown and is thought to be responsible for long distance spread of the disease, but it is not found in the northern United States.

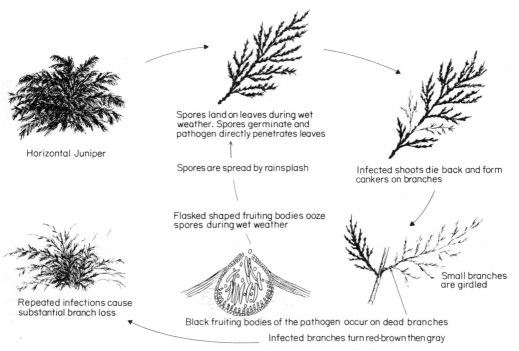

Horizontal Juniper

Spores land on leaves during wet weather. Spores germinate and pathogen directly penetrates leaves

Spores are spread by rainsplash

Infected shoots die back and form cankers on branches

Flasked shaped fruiting bodies ooze spores during wet weather

Small branches are girdled

Repeated infections cause substantial branch loss

Black fruiting bodies of the pathogen occur on dead branches

Infected branches turn red-brown then gray

Fig. 8.23 Disease cycle of juniper twig blight caused by *Phomopsis juniperovora.*

Fig. 8.24 Brown spot needle blight on Scots pine being inspected by Dr. Darroll D. Skilling. Note defoliation of lower branches. (Photo courtesy of Darroll D. Skilling, USDA, Forest Service, St. Paul, Minnesota.)

Treatment   Disease severity can be minimized by use of the same cultural practices recommended for control of Lophodermium needle cast. Since the times of spore release are much earlier in the season for brown spot than for Lophodermium needle cast, applications of fungicides to control brown spot begin in late spring or early summer. The first application is made when the needles are half grown, and the second application is made 3 to 4 weeks after the first application.

Diplodia Tip Blight

Hosts   Several pine species are susceptible including Austrian, mugho, ponderosa, red, Scots, and white pine. Austrian pine, however, is the most susceptible species.

Symptoms   During the spring, on newly emerging foliage, browning occurs near the base of the needle and progresses toward the tip until the entire needle is killed. Infected needles usually are stunted and turn yellow before death (Fig. 8.25). The infection spreads to other needles on the twig, and the entire current year's foliage on an infected shoot is usually killed. The twigs are also infected and are killed back to the next branch whorl.

This tip blight disease usually begins with the lower branches and progresses up the tree. Many lateral shoots begin to form behind killed branch tips and the appearance of the tree deteriorates. As lateral shoots also become infected entire branches may be killed. After several successive years of infection the lower branches are often killed and the entire tree's growth is stunted. Occasionally severe infections can kill trees. This disease is most severe on older, weaker trees.

Disease Cycle   Diplodia tip blight is caused by the fungus *Diplodia pinea* (Fig. 8.26). This fungus overwinters in immature fruiting bodies on plant parts such as dead needles, twigs, cones, seeds, and bud scales, both on the tree and ground. During the spring, as the new needles emerge, the fruiting bodies mature and ooze spores during wet weather. The spores are spread by rain-splash to the current year's needles and cause infections. Most infections occur in mid to late spring when the needles are less than half grown. The pathogen grows extensively through the leaf tissue, then attacks adjacent needles, and eventually spreads to the shoot. Wounds such as those caused by shearing or pruning can also serve as locations for infection on the shoots or twigs. Later in the season fruiting bodies are formed on dead tissue shortly after it is killed, but mature needles are not infected.

Fig. 8.25   Diplodia tip blight on red pine. Note stunted needles on tips of infected branches compared to longer normal needles from the previous year.

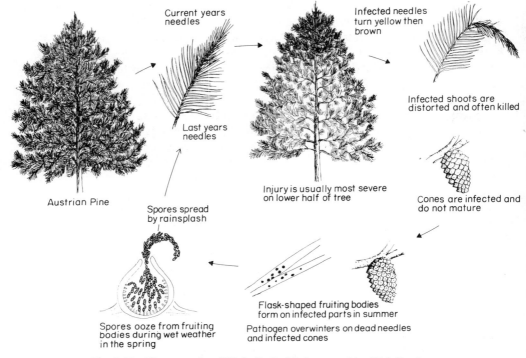

Fig. 8.26 Disease cycle of Diplodia tip blight caused by *Diplodia pinea*.

Treatment   The severity of Diplodia tip blight can be minimized by sanitation and cultural practices to maintain vigor. Infected twigs, branches, and cones should all be removed in dry weather. This will decrease the sources of inoculum and minimize new infections. Trees infected the previous year should be fertilized with balanced fertilizers in late fall or in spring, prior to bud break, and watered during dry periods. It is important to keep infected trees as vigorous as possible since this disease is often most severe on trees that are old or are in poor health.

Effective control of tip blight often requires applications of fungicides in the spring. One application should be made during bud elongation and another in 7 days. Applications made after this period are not effective.

Other Leaf Diseases of Evergreens

There are many leaf diseases of evergreens that have not been discussed. However, many of these diseases such, as Dothistroma needle blight, Elytroderma disease, Rhabdocline needle cast, and Rhizosphaera needle blight have symptoms, life cycles, and controls similar to those of the diseases that have already been discussed. It is also inevitable that anyone working with evergreens will encounter unfamiliar leaf diseases caused by fungi. Since the life cycles of many of these fungus pathogens are similar, the principle of prevention and control are often the same. Therefore, steps taken to minimize disease severity and to protect foliage from the major diseases presented will also protect against most other infectious leaf diseases.

Fig. 8.27 Some insect and mite conditions that mimic leaf diseases. (A) Bladder galls on silver maple caused by mites. (B) Ocellate galls on red maple caused by a midge. (C) Gouty vein on sugar maple caused by a midge. (D) Holly leaf miner. (Photos A and B courtesy of George N. Agrios, and photos C and D courtesy of William B. Becker, University of Massachusetts, Amherst.)

INSECT CONDITIONS COMMONLY CONFUSED WITH DISEASES

Many species of insects and mites feed on hardwood and evergreen leaves and cause symptoms on the leaves that mimic diseases (Fig. 8.27). Those with piercing and sucking mouth parts such as aphids, white flies, and spider mites often cause small discolored or necrotic spots or flecks on leaves.

The feeding of many insects also causes galls on leaves. Leaf miners feed on the tissue between the upper and lower epidermis of the leaf, leaving a dead leaf skeleton. Many of these insect or mite conditions are hard to diagnose accurately because the causal organisms are either very small, hidden inside plant tissue, or absent when symptoms are discovered. Close inspection of leaf tissue with the aid of a hand lens is often required to detect mites. In many cases evidence of insect activity, such as molted exoskeletons, droppings, dead bodies or body parts or eggs, can also be found. Dissection of leaf tissue with a razor blade or pocket knife may be required to determine whether galls or transparent leaf spots are caused by insects or fungi.

There are also numerous nonpathogenic conditions that may be observed on leaves that can mimic diseases. These will be discussed in detail in Chapter 24, Nonpathogenic Conditions.

LITERATURE CITED

Daughtrey, M. L., C. R. Hibben, and G. W. Hudler. (1986). A guide to dogwood anthracnose. *Am. Nurseryman* **163,** 73–74.
Peckhold, P. C., W. R. Stevenson, and D. H. Scott. (1976). Juniper twig blight. *Purdue Univ. Ext. Publ.* **BP-2-18,** 1–2.
Schoeneweiss, D. F. (1969). Susceptibility of evergreen hosts to the juniper twig blight fungus, Phomopsis juniperovora, under epidemic conditions. *J. Am. Hortic. Soc.* **94,** 609–611.

SUGGESTED REFERENCES

Daughtrey, M. L., and C. R. Hibben. (1983). Lower branch dieback, a new disease of Northeastern dogwoods. *Phytopathology* **73,** 365.
Foster, J. A., W. A. Sinclair, and W. T. Johnson. (1973). Diplodia tip blight of pines. *Cornell Tree Pest Leafl.* **A-7,** 1–3.
Himelick, E. B. (1961). Sycamore anthracnose. *Proc. Int. Shade Tree Conf.* **37,** 136–143.
Lightle, P. C. (1969). Brown-spot needle blight of longleaf pine. *U.S., For. Serv., For. Pest Leafl.* **44,** 1–7.
Neely, D. (1976). Sycamore anthracnose. *J. Arbori.* **2,** 153–157.
Nicholls, T. H. (1976). Control of needle diseases—an overview. *Int. Union. For. Res. Organ. World Congr. Proc. Div. II,* 317–329.
Nicholls, T. H., and D. D. Skilling. (1974). Control of lophodermium needle cast disease in nurseries and Christmas tree plantations. *U.S., For. Serv., Res. Pap. NC* **NC-110,** 1–11.
Peterson, G. W., and D. A. Graham. (1974). Dothistroma needle blight of pines. *U.S., For. Serv., For Pest Leafl.* **143,** 1–5.
Peterson, G. W., and C. S. Hodges, Jr. (1975). Phomopsis blight of junipers. *U.S., For. Serv., For. Pest Leafl.* **154,** 1–5.

Peterson, G. (1981). Control of Diplodia and Dothistroma blights of pines in the urban environment. *J. Arbori.* **7,** 1–5.

Skilling, D. D., and T. H. Nicholls. (1974). Brown spot needle disease—biology and control in Scotch pine plantations. *U.S., For. Serv., Res. Pap. NC* **NC-109,** 1–19.

Walton, G. S. (1986). Association of dogwood borer with the recent decline of dogwood. *J. Arbori.* **12,** 196–198.

9

Stem Diseases

INTRODUCTION

Stem diseases are caused by the formation of lesions (dead areas) on the bark-cambium tissues. The death of these tissues is usually associated with a wound, such as a broken branch or some type of mechanical injury, and extends radially from the wound. This localized lesion is called a canker. Stem diseases occur on the trunk, the branches, and the twigs. Cankers that occur only on shoots and twigs, such as anthracnose or juniper twig blight, could be included, but they were covered in the chapter on leaf diseases because the pathogens that cause them also attack leaves. Many rust fungi also cause cankers but these are treated in a separate chapter because of the unique nature of their life cycles.

Cankers vary considerably in size and shape. The development of a canker on a woody plant represents the host–parasite interaction on a stem often for several years. A typical sequence for canker development is as follows: (1) the pathogen enters the host through a wound, and invades and kills healthy bark, usually during a dormant period; (2) the host attempts to limit pathogen invasion by forming a layer of callus over the edge of the infected tissue; (3) the pathogen invades the callus tissue during the next dormant period; (4) the host forms new callus. Steps 3 and 4 may be repeated each year throughout the life of the host. The form of the resulting canker will be determined by the rate of pathogen movement and the amount of callus formation.

Most cankers can be placed into three groups; target cankers, diffuse cankers, and canker blights. Cankers that are roughly circular and contain abundant callus throughout the canker face and at the margin are called target cankers. The pathogen spread in target cankers is relatively slow and in most cases the tree's radial increase in growth is about the same as the radial growth of the canker. Cankers that are elongated ovals and contain little callus at the margin are termed diffuse cankers. The pathogen spread in diffuse cankers is faster than in target cankers; consequently, little callus is formed in advance of the canker. Diffuse cankers enlarge faster than the radial growth of the tree and often girdle trees after several years' infection. Cankers that are circular to elliptical, but contain little or no callus, and increase rapidly during a single season are termed canker blights. The pathogen spread in canker blights is extremely rapid; therefore, branches and even whole trees are often girdled in a single season.

Most canker fungi are restricted to invasion of bark tissues. However, some fungi have the unique ability to attack both the bark and the underlying xylem. Stem diseases that result from a simultaneous canker and woody decay are called canker rots.

TYPICAL LIFE CYCLE OF STEM DISEASE FUNGI

The fungi that cause cankers are usually facultative parasites. They are often found growing saprophytically on the bark of living trees and on dead branches both on the tree and on the ground. A wound is the key event in canker formation. The most common wound site is a branch stub, although mechanical injuries are also common sites for cankers on shade trees. Given a wound and a host in susceptible condition, canker fungi can invade healthy bark and produce a canker. Canker fungi usually produce fruiting bodies on recently killed bark. The spores from these fruiting bodies, as well as those from the same fungi growing saprophytically, serve as inoculum for new infections. Often two types of spores are produced: (1) forcibly discharged airborne spores capable of traveling long distances; and (2) spores that ooze from fruiting bodies and are moved locally by rainsplash.

TREATMENT OF STEM DISEASES

Stem diseases can be minimized by avoiding all unnecessary wounds and by employing sound techniques for wound treatment when pruning, cabling, or any other form of essential wounding is performed on trees (see Chapter 13). Control of a stem disease can be achieved only by removal of all infected tissues. Cankers on small to medium-sized branches can be easily removed by pruning. Cankers on the trunk or large branches that cannot be cut without destroying the value of a shade tree can be removed only by surgical excision of infected bark. Surgical excision is similar to wound treatment around trunk wounds (see Fig. 13.8), except that additional healthy bark is sometimes removed all around fast growing cankers to ensure that all infected tissues are removed. Host vigor has been found to be a contributing factor in the susceptibility of trees to many stem diseases. In general, trees in poor vigor cannot heal wounds and prevent the invasion of canker fungi as well as trees that have abundant moisture and balanced soil nutrients. Therefore, following therapeutic treatments for stem diseases, regular watering and applications of balanced fertilizers are recommended to prevent new infections.

TARGET CANKERS

Nectria Canker

Hosts A large number of deciduous trees are susceptible to this disease, including apple, basswood, birches, black walnut, elm, locusts, maples, and oaks.

Symptoms Oval to elliptical cankers develop around a branch stub or mechanical wound. The canker face may be covered with bark as is common in birches (Fig. 9.1) or it may be exposed dead wood (Fig. 9.2). Inspection of the exposed wood will reveal many concentric callus ridges centering on the wound site, usually a broken branch. Small red fruiting bodies are usually found at the

Fig. 9.1 Nectria canker on sugar maple. (Photo courtesy of Alex L. Shigo, USDA Forest Service, Durham, New Hampshire.)

Fig. 9.2 Nectria cankers on black birch.

margin of the canker that can be easily seen with a hand lens. The concentric callus ridges account for the target appearance of the canker. Since the girth of the tree and the canker diameter are increasing at about the same rate, the tree is rarely girdled in this canker disease. Occasionally, several cankers can occur in close proximity, come together, and girdle the tree. Nectria cankers weaken a tree structurally by decreasing flexibility in the wind; consequently, tree breakage at the canker site is common (Fig. 9.3).

Disease Cycle Nectria canker is caused by the fungus *Nectria galligena* (Fig. 9.4), however, other species of *Nectria* are sometimes known to produce similar cankers. The fungus overwinters as fruiting bodies or vegetative mycelium at the edge of the canker or on the bark as a saprophyte. During the spring fruiting bodies produce spores that are blown or splashed to branch stubs or other wounds. The fungus grows in the bark tissue of the host, causes a canker, and produces small creamy-white fruiting bodies. These spores ooze during wet weather and are spread by rainsplash. These fruiting bodies, however, are not easy to detect and are often overlooked. The next growing season small, red, lemon-shaped fruiting bodies are also produced near the edge of the canker (Fig. 9.5). Windblown spores are forcibly ejected by these fruiting bodies primarily in the summer and fall. Both types of spores serve as inoculum in succeeding years to create new infections.

Treatment Nectria cankers can be minimized by the following cultural practices: pruning should be avoided during wet weather; all unnecessary wounding

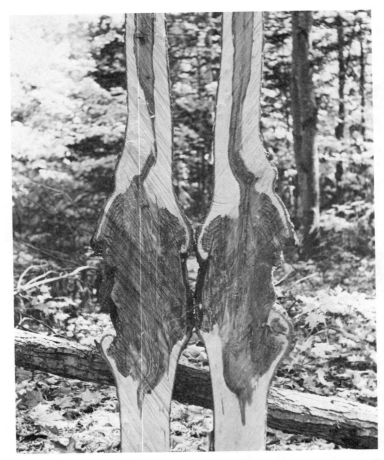

Fig. 9.3 Cross section through a Nectria canker on sugar maple. Extensive bark killing and internal discoloration often cause decreased flexibility resulting in tree breakage at the canker. (Photo courtesy of Alex L. Shigo, USDA Forest Service.)

should be avoided; and rapid healing of wounds should be encouraged by regular fertilization and watering. Nectria cankers can be controlled by pruning out branch cankers and by surgical excision of cankers on the trunk and main branches. Bark should be removed to the cambium at least 1 inch (2.5 cm) beyond the edge of the dead cankered tissue all around the edge of the canker. Cutting tools should be surface-sterilized after each cut. A 20% solution of household bleach or a 70% alcohol solution is recommended. Healthy callus should develop the next season all around the excised tissues. If any dead tissue occurs at the margin, the canker may have advanced beyond the excision. In this case, the dead areas must be excised back to healthy tissue and observed periodically until healthy callus appears.

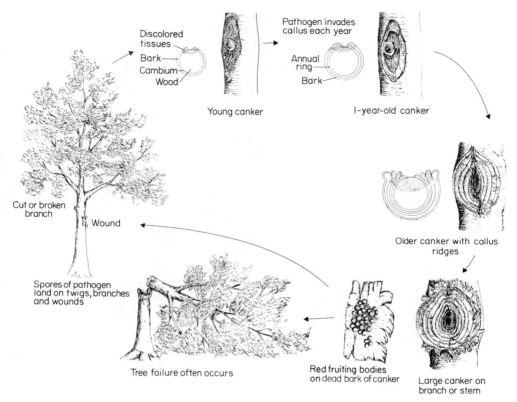

Fig. 9.4 Disease cycle of Nectria canker caused by *Nectria galligena*. [Redrawn from George N. Agrios. (1977). Plant pathology. Academic Press, New York.]

DIFFUSE CANKERS

Dothichiza Canker

Hosts Many species of poplar are susceptible including balsam, black, and Lombardy, and Norway maples and black and eastern cottonwoods, but Lombardy poplar is by far the most susceptible species.

Symptoms Dead branches appear scattered in the upper branches. Closer inspection reveals twig and branch cankers that have girdled the affected stems. Cankers also may be found on larger branches and thé main stem. The cankers are oval sunken areas of bark that may eventually crack and peel off exposing the dead wood beneath. Cankers are often surrounded by callus growth. Small raised black fruiting bodies push through the bark on dead branches shortly after they are killed. In following seasons increased numbers and sizes of

Fig. 9.5 Fruiting bodies of *Nectria* sp. (× 120). (SEM courtesy of Merton F. Brown and H. G. Brotzman, University of Missouri, Columbia.)

branches are girdled (Fig. 9.6). Many trunk sprouts are also produced during this time from the trunk and large branches but the appearance of the tree is decreased markedly by the presence of large numbers of dead branches. The main stem is eventually girdled by cankers, usually in stages from the top down, and the entire tree is progressively killed. Larger trees may take several years to die, from the onset of early symptoms, but small trees are often killed in one or two seasons.

Disease Cycle Dothichiza canker is caused by the fungus *Dothichiza populea* (Fig. 9.7). This fungus overwinters as fruiting bodies on dead branches on the tree and as vegetative mycelium at the margin of cankers. During the spring, spores ooze from the fruiting bodies during wet weather and are splashed to leaves and bark. Infection can occur through the leaves or through bark injuries. Leaf tissue is infected initially, but the pathogen soon moves down the petiole and invades the twig, causing a canker. The expansion of cankers causes stem girdling, and the pathogen then invades the killed distal portions and produces large numbers of fruiting bodies. Spores are oozed during wet weather and can create new infections.

Treatment This disease can be most effectively prevented by avoiding the use of Lombardy poplar. This disease is not usually considered a serious problem in other poplar species. Once a Lombardy poplar is infected, however, there is little that can be done to control this disease. Pruning of infected tissue does not appear to delay the progression of the disease and in some cases has been shown to increase it. Protection against leaf infections with fungicides has been recommended but there is little proof that these applications will have any beneficial effects.

Black Knot

Hosts Sour, sweet, and wild cherries, and cultivated and wild plums are susceptible.

Symptoms Small light-brown swellings occur on the twigs and small branches and continue to enlarge during the growing season. The next spring the swellings turn olive-green but gradually turn darker until they become black. The swellings are velvety in the spring but soon harden and become woodlike during the summer. These hard black swellings are called knots and may range

Fig. 9.6   Dieback of Lombardy poplars from Dothichiza cankers. (Photo courtesy of George N. Agrios, University of Massachusetts, Amherst.)

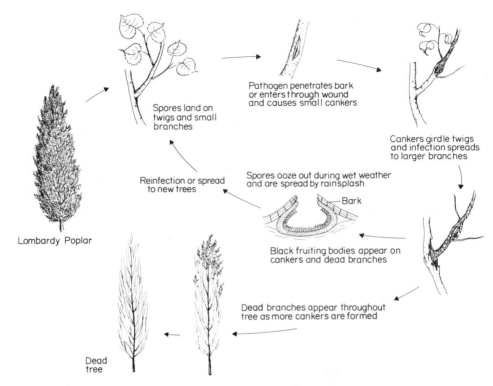

Lombardy Poplar

Spores land on twigs and small branches

Pathogen penetrates bark or enters through wound and causes small cankers

Cankers girdle twigs and infection spreads to larger branches

Reinfection or spread to new trees

Spores ooze out during wet weather and are spread by rainsplash

Bark

Black fruiting bodies appear on cankers and dead branches

Dead branches appear throughout tree as more cankers are formed

Dead tree

Fig. 9.7 Disease cycle of Lombardy poplar canker caused by *Dothichiza populea*.

from 4 to 12 inches (10 to 30 cm) in length or larger (Fig. 9.8). The knots initially cover only part of the branch, but eventually may encircle it. The knots continue to grow until the branch becomes girdled and dies. In many cases large perennial knots develop that disfigure and stunt the tree (Fig. 9.9). Frequently large knots are also invaded by woodboring insects.

Disease Cycle Black knot is caused by the fungus *Dibotryon morbosum* (Fig. 9.10). The fungus overwinters in fruiting bodies on the outside of the knot or as vegetative mycelium in infected twigs and branches. During the spring, spores are forcibly ejected from fruiting bodies during wet weather and cause infections on the current year's twigs. A newly infected branch will swell slightly by fall. The fungus overwinters in the infected branch; the following spring the branch continues to swell and the bark becomes cracked and roughened. During the summer the bark splits and becomes filled with a velvety olive-green material. Summer spores are produced at this time but their role in the disease cycle is considered minor. Lack of extended wet periods during midsummer may account for the general lack of infection by summer spores. The swellings continue to enlarge, turn black, and harden into "knots" by fall. Fruiting bodies are formed on the outside of the knots during fall and winter. The disease cycle takes 2 years to complete.

Treatment Susceptible trees can be protected from infection by applications of fungicides during periods of spore release. The first application should be just before bud break, followed by two more applications at 2-week intervals. Control of infected trees can be achieved by pruning all infected branches at least 4 inches (10 cm) below the knot. Pruning should be done in late fall or early spring before bud break. Knots on large limbs may be surgically excised if they do not extend all around the branch.

Cytospora Canker

Hosts Cytospora cankers occur on a large number of deciduous trees and shrubs as well as a few species of conifers. Blue spruce, maples, mountain ash, Norway spruce, poplars, and willows are the most common hosts.

Fig. 9.8 Black cherry branch heavily infected with black knot cankers.

Fig. 9.9 Black cherry with numerous perennial black knot cankers.

Symptoms This disease exhibits different symptoms on deciduous and on coniferous hosts. On deciduous trees, sunken elongate cankers form on the trunk or branches usually beginning at a branch stub or mechanical wound. The bark on the canker face is often cracked and discolored with a callus ridge at the margin. Small black pimples appear on the dead bark on the canker face. These pimples exude long thin threads during wet weather (Fig. 9.11). Small branches and twigs are rapidly girdled while larger branches and the main stem may take several years to become girdled. On conifers, the lowest branches are killed and then branch killing progresses up the tree (Fig. 9.12). Dying branches first exhibit yellow-green then purple foliage and later cast their needles. A large amount of resin flows from infected branches, which coats the entire surface of the bark around the cankers (Fig. 9.13) and drips onto lower branches. Branch cankers are slightly swollen areas on the branch (Fig. 9.14) but can be identified most easily by their coating of resin. Cutting into the bark, however, will reveal brown areas of killed bark and cambium. Small black pimples also occur on the cankered bark of coniferous trees, but their detection may require close inspection with a hand lens. Thin threads are also produced from these pimples during wet weather. Trunk cankers sometimes occur, especially in blue spruces, which can girdle the entire tree.

Disease Cycle Cytospora canker is caused by two species of fungi in the genus *Cytospora*; *C. chrysosperma* causes cankers on hardwoods while *C. Kunzei* causes cankers on conifers. These fungi overwinter as fruiting bodies on bark and as vegetative mycelium in cankers. During wet weather in the spring fruiting bodies, which appear like small black pimples on the bark, exude thin threads of spores that are washed to other branches by rainsplash. These spores can also be transported by insects, pruning tools, and clothing. If the spores land on a wound they can cause an infection. A canker will eventually develop as the fungus invades healthy bark. Fruiting bodies will form on the dead bark of the canker face and also on dead branches girdled by cankers. Newly formed fruiting bodies will exude spores during wet weather into late fall.

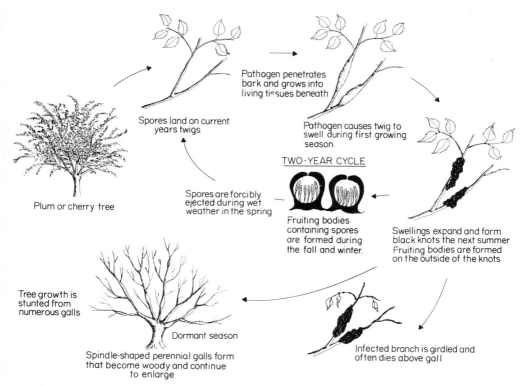

Fig. 9.10 Disease cycle of black knot of plum and cherry caused by *Dibotryon morbosum*.

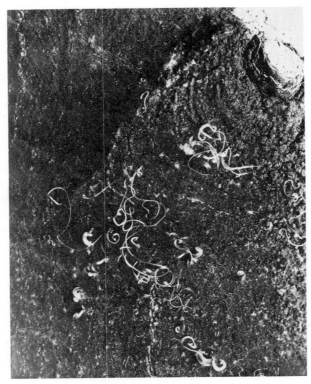

Fig. 9.11 Spore threads or tendrils of *Cytospora* sp. on bark at canker face (× 3). (Photo courtesy of Shade Tree Laboratories, University of Massachusetts, Amherst.)

Treatment Incidence of Cytospora canker can be minimized by keeping susceptible trees as vigorous as possible, since this disease is most severe on trees of low vigor. Dead and dying branches should be removed but pruning should only be done in dry weather. All unnecessary wounding should be avoided. Existing cankers can be removed by pruning or surgical excision. Excision of healthy bark should be made at least 2 inches (5 cm) beyond the canker margin. Both pruning and excision should only be done during dry weather and all cutting tools should be disinfected after each cut. Protective fungicide applications are sometimes recommended following removal of cankers; however, there is no evidence that they offer any protection against future infections.

Fig. 9.12 (A) Disease progression of Cytospora canker on blue spruce. (B) Healthy blue spruce. (C) Blue spruce severely infected with Cytospora cankers. (Photo B courtesy of Shade Tree Laboratories, University of Massachusetts, Amherst.)

A

Spores of pathogen land on wounds in branches near soil during wet weather

Pathogen causes cankers that eventually girdles branches. Spores are spread to other branches

Pathogen progressively kills branches up tree

B

Fig. 9.12   (*continued*)

CANKER BLIGHT

Chestnut Blight

Hosts Allegheny chinkapin, American chestnut, and European chestnut are highly susceptible to chestnut blight, and Japanese and Chinese chestnuts are resistant. The pathogen also causes diffuse cankers on eastern chinkapin, live oak, and post oak.

Symptoms Cankers form on the trunk and branches, beginning from a broken branch or site of mechanical injury (Fig. 9.15). The cankers develop rapidly and eventually girdle the infected trunk or branch (Fig. 9.16). The cankered bark is initially sunken and turns red-orange, but it later becomes swollen and cracked. Numerous orange pustules develop on the dead bark of the canker face and on dead bark of killed stems beyond the canker. Mycelial fans are sometimes produced beneath the bark on the canker face. During wet weather thin threads are exuded from the orange pustules.

Disease Cycle   Chestnut blight is caused by the fungus *Endothia parasitica*. This fungus overwinters as fruiting bodies on dead bark and as vegetative mycelium in the canker. In the spring the pustules, which contain two types of fruiting bodies, discharge their spores during wet weather. One type of fruiting body oozes in thin threads of spores that are spread by rainsplash, while the other type forcibly ejects windblown spores. If the spores contact a wound, an infection can occur and the fungus will invade the healthy bark. After the bark is killed pustules containing both types of fruiting bodies are formed within 2 months. Pustules can discharge spores for several years after the initial infection.

Treatment   Chestnut blight is very difficult to control in highly susceptible species. The pathogen was introduced into Europe and North America from Asia at the beginning of this century, and in the United States it has virtually caused the death of the entire American chestnut forest. All that remain of this

Small cankers on branch cause resin flow

Resin flow

Fig. 9.13   Close-ups of Cytospora cankers on conifers.

Fig. 9.14 Cytospora cankers on spruce branches. Note abundant resin around cankers and the swollen area at the canker (right). (Photo courtesy of Shade Tree Laboratories, University of Massachusetts, Amherst.)

once dominant forest species are isolated individuals and stump sprouts. It is not certain whether large, isolated American chestnuts, free of symptoms, are resistant or merely escape. Since the roots are not killed by this pathogen, stump sprouts are formed after the tops are killed. Most of the sprouts are usually infected and killed before they reach 4 inches (10 cm) diameter; however, larger diameter sprouts sometimes occur. The parent stumps of many chestnut trees have completely decomposed, and it is often difficult to determine whether a tree is a stump sprout or a new tree that originated from seed. Many homeowners, therefore, feel that they may have a disease-resistant American chestnut on their property, especially if it reaches sufficient size to produce seed. Most of these trees, however, are stump sprouts and will eventually develop cankers and die.

Most of the attempts to control chestnut blight have centered on the development of disease-resistant varieties. Some of these varieties have been developed, but unfortunately none of these has the same growth characteristics as the

Fig. 9.15   Chestnut blight cankers on American chestnut. (Photos courtesy of David M. Sylvia, University of Massachusetts, Amherst.)

American chestnut. Resistant varieties of American chestnut as well as the resistant oriental species of chestnut, however, are satisfactory for use as shade trees and for seed production. Anagnostakis and Jaynes (1973) reported success in controlling active chestnut blight cankers with the use of "hypovirulent" strains of the pathogen. The hypovirulent strain does not cause the blight disease but instead outcompetes and replaces the pathogen, causing the cankers to stop enlarging and allowing callus eventually to close over the canker. Whether the hypovirulent strain can be used as an effective control measure for chestnut blight in arboricultural practice is uncertain, but use of naturally occurring antagonistic microorganisms from soil to eradicate chestnut blight cankers was reported by Weidlich (1979). He applied a "mudpack" compress of forest soil to the active canker, wrapped the treated area with a polyethylene bag, and secured it in place with tape. The treated cankers stopped advancing, and within 2 months after treatment the host tree began to produce callus growth around the canker edges.

OTHER CANKER DISEASES

The examples of canker diseases covered in this chapter represent only a brief cross section from the major categories of target cankers, diffuse cankers, and

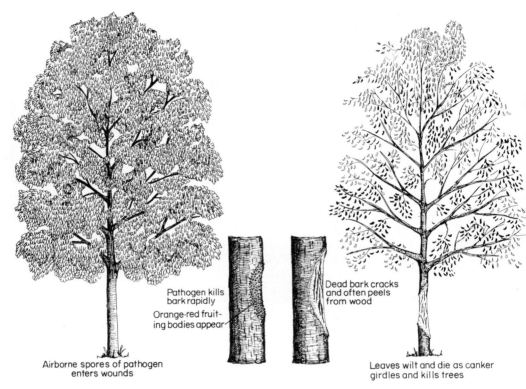

Pathogen kills bark rapidly

Orange-red fruiting bodies appear

Airborne spores of pathogen enters wounds

Dead bark cracks and often peels from wood

Leaves wilt and die as canker girdles and kills trees

Fig. 9.16 Disease progression of chestnut blight caused by *Endothia parasitica*.

canker blights. As in leaf diseases, there are a large number of other canker diseases that attack trees and one can expect to find at least one stem disease for each tree species. Most other canker diseases, however, can be placed into one of these three major categories of cankers. In other canker diseases branch stubs and mechanical wounds are also the principal sites for infection and the disease cycles of most cankers are similar. Therefore, the principles of treatment given for these canker diseases can also be applied to unfamiliar canker diseases of shade trees.

CANKER-ROTS

Canker-rots are important stem diseases of shade trees because, in addition to forming a bark canker, the wood inside the tree is also attacked. Trees infected with canker-rot fungi may have only small cankers but considerable internal defect may exist (Fig. 9.17). Infected trees are often structurally unstable and may break off at the canker face. Therefore, canker-rots are considered one of the major categories of hazard trees (see Chapter 26, Living Hazard Trees).

The disease cycle and symptoms of canker-rots are similar to those of cankers. Branch stubs and mechanical wounds are the most common entry point for the pathogen. The fungi that cause canker-rots, however, are all wood-decay fungi. In some cases fruiting bodies are formed at the canker face, but in others only sterile mycelial plugs form while the tree is still alive. Canker-rots frequently result in the death of the tree either by canker girdling of the main stem or by trunk breakage from extensive internal decay. Some of the most common canker rots are Glomeratus, Hispidus, Obliqua, and Strumella cankers.

Little can be done to control a canker-rot disease. The major cause of concern with canker-rots in shade trees is the potential hazard that an infected tree poses

Fig. 9.17 Irpex canker on willow oak. (A) External view; (B) cross section exhibiting extensive internal decay behind the canker. (Photo courtesy of USDA Forest Service.)

to life and property. If an infected tree is still structurally sound a regular program of fertilization and watering may enable a tree to maintain vigor and slow the progress of the disease for many years. In most cases, however, it is safest to remove all trees infected with canker-rots.

Glomeratus Canker

Glomeratus canker is caused by the fungus *Polyporus glomeratus*. The fungus infects branch stubs of maples, primarily red and sugar, and beech, and produces a small sterile black fungus plug (Fig. 9.18). This plug prevents wound

Fig. 9.18 Longitudinal section through the fungus plug of *Polyporus glomeratus*. (Photo courtesy of Alex L. Shigo, USDA Forest Service, Durham, New Hampshire.)

Fig. 9.19 Exaggerated branch stubs and spindle-shaped callus associated with Glomeratus canker. (Photo A courtesy of Alex L. Shigo, USDA Forest Service, Durham, New Hampshire.)

closure and, through pressure against the callus, allows the fungus continually to invade newly formed tissue. From the outside, the tree may appear to have extremely exaggerated branch stubs with spindle-shaped callus above and below the stub (Fig 9.19). Since the fungus is able to create a new wound each year, it creates a progressively larger column of decay, which may extend almost throughout the entire cross section of the stem and many feet (several meters) above and below the mycelial plug. This disease can be detected by cutting into the swollen knot; the presence of brown-black, cinderlike material is a positive indicator. After the tree dies or is blown over a fruiting body will form.

Hispidus Canker

Hispidus canker is caused by the fungus *Polyporus hispidus*. Dead branch stubs serve as the major entry point of this fungus. Ash, hickory, mulberry, oak, walnut, and willow are commonly attacked. The pathogen invades the healthy bark and produces an elongate sunken canker; decay forms in the wood beneath

Fig. 9.20 Hispidus canker on Nuttall oak. Note large annual fruiting body commonly associated with this disease. (Photo courtesy of USDA, Forest Service.)

the canker and a short distance above and below it. Large bracket-shaped fruiting bodies, yellow-brown to red-brown on the upper surface, are formed during late summer to early winter (Fig. 9.20). New fruiting bodies are formed each year, while old fruiting bodies dry and harden, and eventually fall from the tree.

Obliqua Canker

Obliqua canker is caused by the fungus *Poria obliqua* and commonly occurs on birches. Branch stubs again serve as the major entry point for the fungus, but it may also enter through mechanical wounds. This fungus forms a large black sterile plug from its mycelium, similar to that produced by *P. glomeratus* (Fig.

9.21). *Poria obliqua,* however, produces a much larger plug than found in Glomeratus canker, which appears similar to a large piece of coal attached to the tree. The interior of the plug is yellow-brown. As in Glomeratus canker, the fungus plug prevents wound closure, and through pressure on the bark, creates new wounds each year that are invaded by the fungus. Beneath the plug, decay may extend through most of the cross section of the stem and several feet (approximately 1 m) above and below the plug (Fig. 9.22). In advanced cases the callus formation around the expanding plug causes the entire trunk to swell and appear like a bowling pin in the area of the plug. Obliqua canker is similar to

Fig. 9.21 Obliqua canker on yellow birch. Note large sterile fungus plug of *Poria obliqua* on canker face. (Photo courtesy of Alex L. Shigo, USDA Forest Service, Durham, New Hampshire.)

Fig. 9.22 Longitudinal section through Obliqua canker. Note extensive internal decay above and below fungus plug. (Photo courtesy of Alex L. Shigo, USDA Forest Service, Durham, New Hampshire.)

Spiculosa canker, a canker-rot caused by the fungus *Poria spiculosa*. This pathogen also produces fungus plugs on infected branch stubs. Spiculosa canker occurs on oaks and occasionally on hickories and honey locust.

Strumella Canker

Strumella canker is caused by the fungus *Strumella coryneoidia* and occurs on basswood, beech, red maple, oak, and shagbark hickory. Tree species in the red oak group, however, are the most susceptible. This fungus usually enters

Fig. 9.23 Strumella canker on red oak. Note ridges caused by alternate bark killing by the fungus and callus growth by the host. Also note sucker growth below canker.

through broken branch stubs, kills the surrounding healthy bark, and produces a sunken elongate canker (Fig. 9.23). The canker face may appear somewhat target shaped from a pattern of alternate killing of bark and callus formation over several years. The bark over the canker face may remain intact but, in many cases, some or most of the bark falls off and the decayed wood beneath is exposed. Small brown to black fruiting bodies are often produced on the canker face. Trunks of infected trees frequently become distorted by excessive callus formation around the canker (Fig. 9.24). Sprouts often form on the trunk below the canker and the top of the tree may die from girdling. However, many trees break at the canker face before death occurs from girdling.

Fig. 9.24 Trunk distortion caused by excessive callus formation around a Strumella canker on red oak.

LITERATURE CITED

Anagnostakis, S. L., and R. A. Jaynes. (1973). Chestnut blight control: Use of hypovirulent cultures. *Plant Dis. Rep.* **57,** 225–226.
Weidlich, W. H. (1979). A preliminary report on a method of biological control of the chestnut blight not involving the use of a hypovirulent stain of Endothia parasitica. Proc. American Chestnut Symposium. West Virginia University Press, Morgantown, 79–80. 122p.

SUGGESTED REFERENCES

Anderson, R. L., and G. W. Anderson. (1969). Hypoxylon canker of aspen. *U.S., For. Serv., For. Pest Leafl.* **6,** 1–6.

Anonymous. (1965). Chestnut blight and resistant chestnuts. *U.S., Dep. Agric., Agric. Res. Serv., Bull.* **2068,** 1–21.

Diller, J. D. (1965). Chestnut blight. *U.S., For Serv., For. Pest Leafl.* **94,** 1–7.

French, D. W. (1961). Cytospora canker in Minnesota. *Proc. Int. Shade Tree Conf.* **37,** 126–128.

Hinds, T. E., and R. G. Krebill. (1975). Wounds and canker diseases on western aspen. *U.S., For. Serv., For. Pest Leafl.* **152,** 1–9.

Lightle, P. C., and J. H. Thompson. (1973). Atropellis canker of pines. *U.S., For. Serv., For. Pest Leafl.* **138,** 1–6.

McCracken, F. I., and E. R. Toole. (1974). Canker-rots in southern hardwoods. *U.S., For. Serv., For. Pest Leafl.* **33,** 1–4.

Shigo, A. L. (1969). How Poria obliqua and Polyporus glomeratus incide cankers. *Phytopathology* **59,** 1164–1165.

10

Root Diseases

INTRODUCTION

The health of the root system is perhaps the most important factor in the total health of a tree. However, most often it is first the leaves, next the stem, and last, if at all, the roots that are examined when the health of a plant is judged. The trunk, branches, and leaves can only grow at the rate permitted by the condition and growth of the root system. The initial symptoms of most root diseases are similar, slow to rapid decline and death of the tree, usually beginning in the upper branches. Unfortunately, a large part of the root system is often killed before any obvious symptoms appear and, consequently, it is often difficult or impossible to control root diseases after appearance of decline symptoms. Root diseases may at first appear mysterious because our knowledge of roots and root growth lags behind our knowledge of other parts of the tree's anatomy. The growth and function of roots, therefore, will be discussed briefly before specific root diseases are examined.

GROWTH OF TREE ROOTS

Until recently, little was known about how tree roots grow in soil. Many people thought that most roots grew sharply downward after leaving the trunk. Careful examination of roots around standing trees indicated that most roots remain in the top 18 inches (45 cm) of soil and that a large number of small roots exist in the top few inches (several cm) of soil. It is, therefore, easy to see why the tree's roots are so easily injured during even minor disturbances in the soil around them and also why trees are so sensitive to the condition of the upper soil layers.

MYCORRHIZAE

It was also thought that roots absorb soil nutrients and water primarily through root hairs on the feeder roots; however, recent evidence indicates that most absorption occurs through feeder roots that are infected by beneficial fungi. These fungi form structures with the feeder roots called mycorrhizae. Practically all forest trees and most other plants are now known to form mycorrhizae.

The tree, therefore, is actually a dual organism that is part plant and part root-inhabiting fungus. The cortical tissues of the young roots are invaded by these specialized beneficial fungi, which enter into a relationship that benefits both organisms. The formation of mycorrhizae aids water and mineral absorption for the tree, and the fungus in turn receives needed organic compounds from its association with the tree.

There are two major categories of mycorrhizae: (1) ectomycorrhizae and (2) endomycorrhizae. Ectomycorrhizae occur on many tree species, including beech, birch, eucalyptus, fir, hickory, larch, pine, oak, and spruce, and have been studied extensively. Most of what we know about mycorrhizae is based upon this group. They are easy to recognize due to the swollen appearance of the feeder roots (Fig. 10.1). Ectomycorrhizae form a fungus covering around the outside of the feeder root, called the mantle, as well as many hyphal strands between the cells in the cortex, called the Hartig net. Both the mantle and Hartig net provide protection for the roots from pathogenic root fungi. Endomycor-

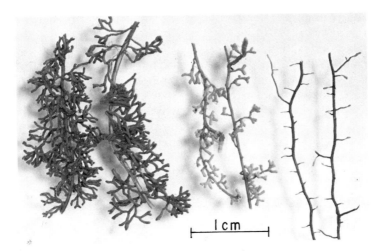

Fig. 10.1 Ectomycorrhizae on three loblolly pine seedlings. Note abundant branching and swollen feeder roots in the two samples (left, center). Roots at far right are similar to nonmycorrhizal roots. (Photo courtesy of Don H. Marx, USDA, Forest Service, Athens, Georgia.)

rhizae also occur on many trees, such as almond, apple, avocado, azalea, box-wood, camellia, citrus, dogwood, holly, maple, palms, plum and sycamore, but have been studied much less than ectomycorrhizae. The colonization of endomycorrhizae usually does not cause any obvious changes in the root although their role in root absorption is thought to be similar to ectomycorrhizae. Endomycorrhizae do not form a mantle or Hartig net but instead form absorbing hyphae, called haustoria, in the living cortical cells of the roots. It is not known whether endomycorrhizae provide any protection against soil pathogens.

TYPICAL LIFE CYCLE OF ROOT DISEASE FUNGI

Many fungi that cause root diseases are facultative parasites that are commonly found growing saprophytically in the soil on organic matter such as dead roots or stumps. From this base living trees are continually attacked. Trees of low vigor are most easily infected, but some fungi are able to infect vigorous trees. Some root disease fungi penetrate directly into healthy roots, but most gain entrance through wounds in the lower trunk, buttress, and roots. Entrance also occurs across points where healthy and infected roots are in contact or are fused. These points, known as root contacts and root grafts, are common when trees of the same species are growing next to each other and their root systems are intermeshed. Once established these fungi progressively kill the roots by girdling. These fungi eventually move into the buttress area and sometimes into the lower trunk. The underlying wood may or may not be decayed, depending on the fungus involved. Infected trees are killed most often by girdling of the lower trunk or buttress but may also fall victim to "wind throw" from extensive decay in the root system.

Most root disease fungi produce specialized structures for survival or movement in the soil. These may be reproductive structures, such as thick-walled or motile spores, or vegetative structures that enable translocation or storage of materials. Some fungi produce windblown spores that are important for long-range disease spread. Since most fungus pathogens of trees' roots are widespread, the local movement of the pathogen in the soil and the movement from one shade tree to an adjacent one are of most concern.

TREATMENT OF ROOT DISEASES

Root diseases can be minimized by keeping trees vigorous and avoiding wounds in the root system and near the base of the tree. Once a tree has become infected it is difficult to control a root disease. A thorough examination of the root system near the trunk is needed to determine the extent of the infection (see Chapter 25, Disease Diagnosis). In severe cases where most of the roots are dead or the trunk is almost girdled, removal is recommended. Where only a few roots are infected, therapy may delay the disease progression almost indefinitely. Since infected trees are often suffering from some previous stress, the vigor of the tree must be restored. Regular watering and fertilization are recommended as well as pruning the crown to balance the root loss. If infected bark occurs on buttress roots and trunk it should be excised to healthy wood.

Replanting in the same area following death from root disease should be preceded by removal of as much of the dead stump and roots as possible. Soil that is infested with the pathogen should be replaced or fumigated. Newly planted trees, in general, are quite susceptible to root diseases. Soil fumigation or trenching around the infected tree should also be done to protect the adjacent susceptible woody plants from the spread of the pathogen in the soil.

CATEGORIES OF ROOT DISEASES

Most fungi that cause root diseases can be placed into one of two categories: (1) fungi that primarily kill small roots and do not decay the large roots, and (2) fungi that cause decay primarily in the larger roots. Two diseases in the first category, Phytophthora root rots and cotton or Texas root rot, and two diseases in the second category, shoestring root rot and Lucidus or Ganoderma root rot, will be discussed in detail.

SOME EXAMPLES OF ROOT DISEASES

Phytophthora Root Rots

Hosts   Phytophthora root rots have a wide host range of trees and shrubs and are found all over the world.

Symptoms   The initial symptoms are usually similar to the onset of a general decline. In conifers, the current year's foliage turns yellow-green and is stunted when compared to the previous year's needles. In successive years the entire foliage appears yellow and sparse with stunted needles in small, upturned tufts at the tips of the branches. In hardwoods, small, light-green leaves are formed that produce fall coloration and drop prematurely. In successive years the foliage becomes more chlorotic and may curl and scorch during the summer. In late stages on both conifers and hardwoods, dead branches occur in the crown and sprouts may occur on the trunk. Twig and trunk growth decrease dramatically during the disease progression. Initially, small roots are killed in increasing numbers and then brown to black lesions occur on large roots. Later, lesions occur on large roots and the buttress. In some Phytophthora root rots, however, the pathogen attacks only the small roots and large roots never develop lesions. In either case, infected trees may decline slowly over several seasons or be killed rapidly in one or two seasons. The aboveground symptoms, however, are typ-

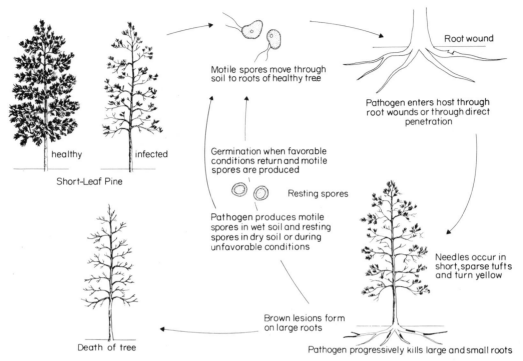

Fig. 10.2 Disease cycle of littleleaf disease of pine caused by *Phytophthora cinnamomi*.

ical of a large number of root diseases. The appearance of the foliage and growth rate of trees are similarly affected whenever the roots are infected or injured.

Disease Cycle Phytophthora root rots are caused by several species of fungi in the genus *Phytophthora*, including *P. cactorum, P. cinnamomi*, and *P. citricola* (Fig. 10.2). These fungi overwinter in soil as thick-walled resting spores or as vegetative mycelium in infected roots and plant debris. During the spring the spores germinate and can directly penetrate into epidermal cells or small roots or enter through wounds. Under wet soil conditions these fungi can also produce motile spores that are propelled by tiny flagella. These motile spores can only travel short distances in the soil but are produced in large numbers and their production, therefore, can accelerate disease development on a tree. Infected rootlets are killed in increasing numbers as more roots become infected. Larger roots are progressively infected and the tree eventually dies. Many of these pathogens also have the ability to cause cankers on the trunk and branches. During the spring, summer, and fall the fungus may produce both motile and nonmotile spores. In cold weather or extended dry periods thick-walled resting spores are produced.

Treatment   The incidence of Phytophthora root rots can be minimized by keeping the trees as vigorous as possible. When the roots of vigorous trees are attacked by *Phytophthora* spp. they are often able to stop the invasion of the pathogen and produce new feeder roots. The presence of abundant mycorrhizae on roots of vigorous trees has also been related to resistance to these diseases. Trees under stress or trees with poor mycorrhizal development are not able to stop the invasion of *Phytophthora* spp. and are killed. Phytophthora root rots are most prevalent in areas with high soil moisture and low soil fertility.

Control of mildly infected trees sometimes can be achieved for many years by applications of balanced fertilizer. In cases where the pathogen has only invaded a small amount of the main roots or buttress, infected bark can be excised back to healthy tissue as is done for cankers and wounds (see Chapter 9, Stem Diseases; Chapter 13, Wound Diseases). Before planting a tree where another tree has died from a Phytophthora root rot, as much of the root system as possible should be removed and the soil should be fumigated. This practice is also recommended even if new plantings are not planned but adjacent trees are in root contact with the infected tree and need protection. Most soil pathogens can persist on dead roots and other organic matter for many years after the host has been killed. These pathogens should be eradicated before new trees are planted in the same location. *Phytophthora* spp. have such an extensive host range of woody plants worldwide that every tree and shrub can be considered to be susceptible.

Cotton Root Rot (Texas Root Rot)

Hosts   Cotton root rot attacks a large number of trees and shrubs in the southwestern United States. Trees that are very susceptible include the Chinese elm, cottonwood, pepper tree, and umbrella tree. Moderately susceptible trees and shrubs include euonymus, fruitless mulberry, jasmine, pecan, and pyracantha. Resistant trees and shrubs include ailanthus, apricot, black walnut, eucalyptus, green ash, Kentucky coffee-tree, oleander, pomegranate, and western soapberry.

Symptoms   The foliage suddenly wilts and dries during the hot summer months. Dead leaves usually remain attached to the tree. Cotton white fungus mats are produced on the soil under an infected tree after a rain or watering. The mats are approximately 3 to 12 inches (7.5 to 30 cm) in diameter and oval or irregular in form. They turn light tan and powdery within 3 or 4 days after they appear. Large numbers of roots are killed by the pathogen in a short period. The severity of infection can be best determined by digging a zone 2 to 3 feet (60 to 90 cm) around the tree. A dead root will appear brown after removal of bark while a living root will be white. Under closer examination with a hand lens the strands of the pathogen can often be seen in bark crevices and in root crotches. Positive identification, however, often requires microscopic examination. The strands remain only a few weeks after a root is killed and then disappear. Most infected trees die the same season unless control measures are applied. Occasionally, trees infected in late summer or fall will appear yellow-green and have sparse foliage the next spring. These trees are usually killed during the following summer.

Disease Cycle Cotton root rot is caused by the fungus *Phymatotrichum omnivorum*. This fungus overwinters as vegetative mycelium in the soil or in infected roots. The fungus is most active during the hot summer months when it grows through the soil and penetrates the roots of susceptible trees. It spreads rapidly by thin strands of mycelium around the root system. Extensive root killing causes moisture stress on the foliage with subsequent wilting. As more roots are killed the tree dies. Spores are not produced but new hosts are infected as the fungus grows rapidly through the soil from infected roots to susceptible plants nearby. The strands also enable *P. omnivorum* to survive long periods in the soil. Therefore, when susceptible trees are replanted they are often infected by this fungus and killed.

Treatment Cotton root rot can be prevented in most cases by selecting only resistant species and by treating the soil to inhibit growth of *P. omnivorum* before a tree is planted. Soil treatment as prescribed by Streets (1974) consists of mixing seasoned manure, sulfur, and ammonium sulfate with the soil to be used for the planting hole. These ingredients are applied in alternate layers beginning at the bottom of the hole, in the following proportions: for a 30 ft^3 (1m^3) planting hole use 6 ft^3 (0.2 m^3) manure, 7.5 lb (3.4 kg) sulfur, and 1.75 lb (0.8 kg) ammonium sulfate. Each layer consists of 2 inches (5 cm) of manure at the bottom with sulfur and ammonium sulfate scattered on top, and covered with 3 inches (7.5 cm) of soil.

Once a tree has become infected, a similar soil treatment can often control the disease or at least prolong the life of the tree. As soon as the first symptoms are recognized, the entire soil area over the roots should be drenched with a solution of ammonium sulfate (1 lb/10 ft^2 or 0.45 kg m^2). Ammonium phosphate may be substituted but ammonium nitrate or any other nitrate fertilizer should not be used. This drench should be soaked into the soil to an approximate depth of 3 feet (1 m). The objectives of the treatment are to stop further root killing by the fungus and to stimulate the growth of the root system. The speed of diagnosis and application of therapy are critical in the effectiveness of this treatment. As root damage increases, the probability of saving the tree decreases.

Approximately 25% of the branches should be removed from a tree showing foliar symptoms, including all wilted branches. Following therapy, infected trees should be watered during dry periods. Sometimes infected trees are subject to windthrow due to a severely damaged root system. Any tree that does not seem firmly anchored should be guyed to prevent windthrow until the newly formed roots are again able to anchor the tree. If the diagnosis is in error and the cause of the symptoms is another root problem, such as moisture stress, underground gas, or nematodes, the treatments do not harm the tree and in most cases are beneficial.

When a tree has been killed from cotton root rot the soil must be free of the pathogen before another tree can be successfully planted in its place. In this case the previous soil treatment may not be sufficient to prevent infection, and soil fumigation is recommended.

Shoestring Root Rot

Hosts   Shoestring root rot affects a wide range of shrubs and trees that are commonly used for ornamentals. It is a serious root disease in shade, forest, and orchard trees all over the world.

Symptoms   The first observable symptoms of shoestring root rot can range from a slow gradual dieback to the sudden death of a tree. Examination of infected roots reveals the presence of mats of fungus mycelium between the bark and the wood (Fig. 10.3). These mats may extend to the buttress area and up the lower trunk. Strands of "shoestrings" composed of fungus mycelium are common on hardwoods and may form between the bark and wood, in decayed wood or around the outside of a root (Fig. 10.4). The dark brown to black shoestrings are called rhizomorphs. In the late summer or early fall honey-colored mushrooms may appear around the base of infected trees (Fig. 10.5). They persist only for a few weeks but the black shriveled remains of the mushrooms can often be found for many months.

Fig. 10.3   Shoestring root rot. Cankers on lower trunks of magnolia (A) and white fungus mats beneath bark of an infected eastern hemlock root (B).

Fig. 10.4 Shoestrings (rhizomorphs) of *Armillaria mellea* under the bark of a recently killed tree (A). Close-ups of rhizomorphs (B). (Photo A courtesy of George N. Agrios, University of Massachusetts, Amherst.)

The root bark is usually killed in advance of the mycelial mats. In conifers, resin flow occurs from infected roots and coats the bark. The wood beneath the mats may be decayed or sound. Decayed wood is moist, bleached, and stringy. Considerable decay of large and small roots, as well as internal decay in the buttress area, is frequently associated with shoestring root rot. Bark killing eventually extends completely around the root buttress or lower trunk and girdles the tree. Many trees are subject to windthrow due to the loss of supporting roots.

Disease Cycle   Shoestring root rot is caused by the fungus *Armillaria mellea* (Fig. 10.6). The fungus overwinters as rhizomorphs or as vegetative mycelium in

Fig. 10.4 (*continued*)

both living and dead trees. During the spring the rhizomorphs and vegetative mycelium resume growth through the soil and infect healthy roots. Rhizomorphs enable the fungus to move from an infected tree to an adjacent healthy one, while vegetative mycelium primarily enables local infection to spread within a tree's root system. After the infected tree is killed, the pathogen lives for many years as a saprophyte. In the late summer or early fall mushrooms produce large quantities of windblown spores. However, since this fungus has already been found in most soils worldwide, it is doubtful that these spores play an important role in the occurrence of this disease.

Treatment   The occurrence of shoestring root rot, in most cases, is related to previous stress on the tree. Since the pathogen exists in most soils, infection is

Fig. 10.5 Honey mushrooms of *A. mellea* at the base of an infected tree. (Photo courtesy of Alex L. Shigo, USDA, Forest Service, Durham, New Hampshire.)

usually related to a lowering of host vigor. The stress may be environmental, such as drought, flooding, or poor drainage; people-caused, such as construction, fill, or chemical injury; or biotic, such as severe or repeated defoliation by insects or leaf diseases. Prevention of shoestring root rot, therefore, depends to a large part on helping the tree avoid stress and remain as vigorous as possible.

Trees already infected with *A. mellea* should be relieved of continuing stress and made as vigorous as possible with regular fertilization and watering. There is no control for shoestring root rot, but, in many cases, the progress of the disease can be postponed almost indefinitely if the tree can be returned to normal vigor. The chances of success will depend on early detection before a large amount of the root system is killed. Severely infected trees constitute a hazard due to loss of physical support and should be removed.

Infected bark on large roots, buttress, or trunk should be excised back to healthy tissue as is done for cankers and wounds (see Chapter 9, Stem Disease; Chapter 13, Wound Diseases). The soil should not be replaced around treated roots until late fall. After excision, exposure of infected roots to air will keep the bark dry and decrease the chances of reinfection. Healthy callus should form around the edges of all excised bark later in the season or the next spring. If callus fails to appear in an area, the condition of the bark should be checked to determine if the pathogen has advanced beyond the point of excision into the healthy tissues. In this case, additional bark must be excised to remove infected tissues. This area should be examined periodically until healthy callus appears. The crown should also be thinned to compensate for loss of roots. Numerous chemical therapeutants have also been applied to trees infected with *A. mellea*; however, the efficacy of these compounds is still not proven. Pawsey and Rahman (1976) present a comprehensive review of research on the attempts to develop chemical controls for shoestring root rot.

When a tree has been killed from shoestring root rot, as much of the old root

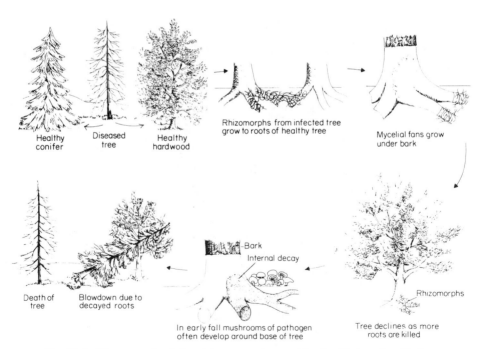

Healthy
conifer

Diseased
tree

Healthy
hardwood

Rhizomorphs from infected tree
grow to roots of healthy tree

Mycelial fans grow
under bark

Death of
tree

Blowdown due to
decayed roots

Bark

Internal decay

In early fall mushrooms of pathogen
often develop around base of tree

Rhizomorphs

Tree declines as more
roots are killed

Fig. 10.6 Disease cycle of shoestring root rot caused by *Armillaria mellea*. [Redrawn from George N. Agrios. (1969). Plant pathology. Academic Press, New York.]

system as possible should be removed before replanting. In addition, soil around the old root system should be fumigated or replaced with fresh soil. The objectives of these treatments are to reduce the amount of *A. mellea* in the soil and to remove any substrate that the fungus can live on.

Lucidus or Ganoderma Root Rot

Hosts A large number of species are commonly attacked including: ash, elm, hackberry, lemon, locust, maple, mimosa, oak, orange, palms, sweet gum,

Fig. 10.7 Crown decline from *Ganoderma* root rot on European beech (A). Close-up (B) showing fruiting body (arrow) of pathogen at soil line. Note death of bark at base of trunk.

Fig. 10.8   Decline in *Reclinata* palm (*Pheonix relinata*) due to *Ganoderma* root rot (A). Close-ups of fruiting body of pathogen (B and C).

tupelo, and willow. Red and Norway maples are the most commonly attacked species in the northeast while mimosa is very susceptible to attack in the southern United States.

Symptoms   The first indications of this disease are often rapid decline and death of the infected tree (Figs. 10.7 and 10.8). Large red-brown, shelf-shaped fruiting bodies form at the base of the infected tree, usually just above the soil line and often partially hidden by turf. The fruiting bodies have a shiny or "varnished" upper surface and a white under surface. Infected roots are decayed and appear bleached white with scattered black specks. Severely infected trees have lost most of their anchoring roots and, therefore, are subject to windthrow. Most infected trees die soon after initial symptom develop; however, some may decline slowly over several years.

Disease Cycle   Lucidus or Ganoderma root rot is caused by the fungus *Polyporus lucidus,* which is also known as *Ganoderma lucidum.* This fungus overwinters as vegetative mycelium in the roots of infected trees or in dead organic matter in the soil. During the spring, the fungus resumes activity and can attack healthy trees through wounds on the roots and buttress area. The pathogen can also spread from one tree to the next via root contacts and root grafts. Fruiting bodies often do not form until several years after the initial infection. The fungus has often caused decay in a large amount of the lower trunk, buttress, and root system before fruiting bodies or decline symptoms appear. Windblown spores are produced in large numbers by the fruiting bodies. These spores can contact wounds near the base of a healthy tree and create infections; however, most infections are thought to occur from pathogen movement through the soil or along infected roots.

Treatment   The best prevention for Lucidus or Ganoderma root rot is to avoid the major sites of infection—trunk and root wounds. Care should be taken to prevent mechanical injury around trees by lawnmowers, construction practices, or other disturbances. This root disease also occurs more commonly on trees stressed by deep planting, addition of fill, or subjected to soil compaction. In the southern United States a major stress factor is detopping of mimosa trees. Larger trees are frequently pruned back to a main stem 6 to 10 feet (1.8 to 3 meters) high to create an abundance of sucker growth. *Polyporus lucidus* often attacks these weakened trees and causes their eventual death. Sometimes this root rot is confused with mimosa wilt (see Chapter 12, Wilt Diseases), but the presence of large fruiting bodies at the base of the trunk and lack of vascular discoloration in the branches allow accurate detection in the field.

Control of this disease in infected trees is difficult because extensive killing of the root system usually occurs before initial symptoms appear. The infected trees are often a hazard to people and property and should be removed as soon as possible. If the disease is detected before many supporting roots are killed the infected tissues can be excised, and the tree can be given other therapy, similar to the treatments recommended for shoestring root rot. Since adjacent healthy trees can become infected from root grafts and root contacts, these should be severed by digging a 3 foot (1 meter) deep trench between an infected tree and any adjacent trees. Once a tree has been killed by this root rot, as much of the root system as possible should be removed, and the soil fumigated or replaced before replanting.

LITERATURE CITED

Pawsey, R. G., and M. A. Rahman. (1976). Chemical control of infection by honey fungus, Armillaria mellea; A review. *J. Arbori.* **2**, 161–169.
Streets, R. B. (1974). Preparing tree holes for root rot control. *Univ. Ariz. Co-op Ext. Pub.* **Q-51**, 1–3.

SUGGESTED REFERENCES

Boyce, J. S., Jr. (1964). Littleleaf diseases of pine. *Proc. Int. Shade Tree Conf.* **40**, 171–177.
Cheo, P. C. (1972). The oak root rot disease and possible means of control. *Proc. Int. Shade Tree Conf.* **48**, 99–101.
Childs, T. W., and E. E. Nelson. (1971). Laminated root rot of Douglas-fir, *U.S., For. Serv., For. Pest Leafl.* **48**, 1–7.
Fisher, R. (1973). Trees grow up by growing down. *Proc. Int. Shade Tree Conf.* **49**, 85–87.
Haeskaylo, E. (1971). Mycorrhizae—Proceedings of the first North American conference on mycorrhizae. *U.S., Dep. Agric., Misc. Publ.* **1189**, 1–255.
Leaphart, C. D. (1963). Armillaria root rot. *U.S., For. Serv., For. Pest Leafl.* **78**, 1–8.
Marx, D. H. (1970). The beneficial relationship between tree roots and mycorrhizal fungi. *Proc. Int. Shade Tree Conf.* **46**, 89a–94a.

Marx, D. H., and W. C. Bryan. (1975). The significance of mycorrhizae to forest trees. *In* Forest soil and forest land management (B. Bernier and C. H. Winget, eds.), pp. 107–117. Presses Univ. Laval, Quebec.

Raabe, R. D. (1969). Soil-borne diseases of shade trees. *Proc. Int. Shade Tree Conf.* **45,** 259–261.

11

Rust Diseases

INTRODUCTION

Rust diseases are caused by a specialized group of fungi called the rust fungi. These fungi are obligate parasites, which means they can only grow on a living host. Rust fungi have several spore stages in their life cycles. Many rust species have five spore stages while others have as few as three spore stages. Most rust fungi that infect trees have spore stages in two completely unrelated hosts. These fungi must move from one host to the other to complete their life cycles. However, some tree rust fungi complete their entire life cycle on one host.

Rust fungi can infect leaves, twigs, branches, and the trunk. These fungi cause spots and blisters on leaves and cankers and galls on branches and the main stem. Rust diseases are often conveniently separated into categories such as canker rusts, gall rusts, and leaf rusts. Canker rusts usually produce diffuse cankers on the trunk or main branches. Gall rusts develop swellings on the trunk and also on a large number of branches. Leaf rusts cause pustules on the foliage of broad-leaved trees and conifers.

TYPICAL LIFE CYCLE OF RUST DISEASE FUNGI

A typical life cycle of rust disease fungi includes five spore stages on two different hosts, although in some species repeating spores are not produced. However, one major difference between life cycles of some tree rust fungi is the overwintering stage. Rust fungi can overwinter either as vegetative mycelium in newly infected leaves and at the edge of active galls and cankers, or they can overwinter as thick-walled resting spores. In the spring thick-walled resting spores germinate and produce windblown spores that infect leaves. Pustules or blisters are formed on the leaves or, in the canker and gall rusts, the fungi move into the branches where spore stages protrude through the bark. These pustules produce spores that are exuded in an ooze that is attractive to insects. The insects move from one pustule to another, feeding on the ooze and carrying spores with them. The insects serve a similar function that of honeybees in the pollination of flowers except that in this case the insects bring together different strains of the fungus and allow it to produce the next spore stage. After mixing of strains, the pathogen produces open blisters on the bark or cluster cups on the bottom of leaves. These form windblown spores that infect the foliage of the other host. The spores penetrate the host and soon produce lesions that are filled with repeating spores. These spores can reinfect the same host species and can repeat the process several times during the early growing season, causing a rapid disease buildup. After several weeks, production of repeating spores ceases and thick-walled resting spores, often in clusters forming a spore horn, are produced from the lesion. The next spore stage, windblown spores that infect foliage, may be produced from these resting spores later during the current season or during the next spring.

TREATMENT OF RUST DISEASES

Gall rusts and canker rusts are usually serious disease problems while leaf rusts are often of minor importance. Gall and canker rusts often occur on the trunk and cause stunting above the infection point. Girdling and/or wind breakage may eventually occur at the infection point as the disease progresses. Leaf rusts cause small infections on foliage, but the foliage usually remains green and does not fall from the tree. Severe infections, however, can cause abscission of foliage. Leaf rusts cause minimal injury to the tree unless most of the foliage is infected or severe infections occur for several successive years. Rust diseases can be controlled by eradication of alternate hosts, by pruning or surgical excision of infected tissues, by protection of foliage from infection with fungicides, and by the use of resistant species. When a rust fungus has two hosts, usually only one is a desirable shade tree. Removal of all undersirable host species in the vicinity will prevent or minimize the occurrence of the disease, because the rust fungi are prevented from completing their life cycles. Alternate hosts, however, must often be removed for at least 1 mile (1.6 km) to achieve complete protection. Foliage can be protected by fungicides, which must be applied during periods of peak spores production of the pathogen when on the alternate host. Many applications are often required at intervals of 7 days or less to achieve protection. Resistant host species that are similar in appearance to highly susceptible species can often be substituted in areas where a rust disease is a persistent problem.

CANKER RUSTS

White Pine Blister Rust

Hosts   White pine blister rust occurs on pine species with five needles in a fascicle, which are called white pines, and on currants and gooseberries and other species of shrubs in the genus *Ribes*. Limber, sugar, western white, and white-barked pines are very susceptible, while eastern white and foxtail pines are moderately susceptible, Macedonian and Swiss stone pines are resistant, and Oriental white and pinon pines are immune. Practically all members of the genus *Ribes* can serve as an alternate host.

Fig. 11.1   White pine blister rust canker on main stem. Note abundant resin on trunk and branches below the canker. (Photo courtesy of USDA, Forest Service.)

Symptoms Swollen areas of bark appear on the infected branches that turn yellow-brown and form cankers. A dead branch, girdled by a canker, is often the first symptom but it is usually overlooked. Resin flows all around the margin of the canker. Cankers advance into the main stem and grow around the trunk (Fig. 11.1). The entire top of the tree will develop sparse, yellow-green foliage, and eventually die as the canker girdles the main stem. Infected trees occasionally break at the canker during high winds (Fig. 11.2). During the late spring or early summer fruiting bodies are formed below the bark and ooze a honey-colored, spore-bearing exudate. The exudate is attractive to insects and birds, which feed on it. Soon after this stage blisters appear on the edge of the canker face. Blistered bark is attractive to squirrels and porcupines who sometimes chew off bark around the edge of the canker (Fig. 11.3). In the summer bright orange pustules form on the leaves of ribes plants, primarily on the underside. Later in the summer and in the early fall, long curled hairs form on the bottom of the leaves.

Disease Cycle White pine blister rust is caused by the fungus *Cronartium ribicola* (Fig. 11.4). This pathogen overwinters as vegetative mycelium in infected needles and in active cankers on white pines. During the spring the fungus resumes growth, and from the needles it eventually grows into the twig and

Fig. 11.2 White pine tree broken off at blister rust canker. (Photo courtesy of USDA, Forest Service.)

Fig. 11.3 Porcupine or squirrel feeding on white pine blister rust canker. (Photo courtesy of USDA, Forest Service.)

adjoining branch. An enlarging canker is formed on the branch, which produces spores that ooze from blisters in late spring or early summer, beginning 2 to 4 years after infection. The spread of these spores allows fungal mating to occur, resulting in the formation of the next spore stage. Open blisters containing this new spore stage are formed 3 to 5 years after infection. These yellow spores are windblown to the leaves of ribes plants where infection takes place. Lesions are formed primarily on the undersides of the ribes leaves and produce a repeating spore stage within a few weeks. During the early summer these spores are able to reinfect other ribes plants. In late summer, long curled spore columns are produced in the lesions that earlier produced the repeating spores. From late summer to early fall these columns produce windblown spores that can infect needles of white pine trees through the stomates.

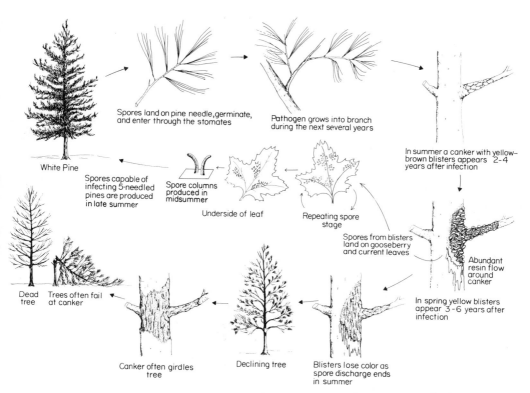

Fig. 11.4 Disease cycle of white pine blister rust caused by *Cronartium ribicola*.

Spores land on pine needle, germinate, and enter through the stomates

Pathogen grows into branch during the next several years

In summer a canker with yellow-brown blisters appears 2–4 years after infection

White Pine

Spores capable of infecting 5-needled pines are produced in late summer

Spore columns produced in midsummer

Underside of leaf

Repeating spore stage

Spores from blisters land on gooseberry and current leaves

Abundant resin flow around canker

Dead tree

Trees often fail at canker

In spring yellow blisters appear 3–6 years after infection

Canker often girdles tree

Declining tree

Blisters lose color as spore discharge ends in summer

Treatment   Currants and gooseberries should be removed in an approximate radius of 900 feet (300 m) from any susceptible white pine tree. This local removal of the alternate host will prevent infection because the thin-walled spores that infect pine needles can only survive transport short distances in the air. Planting highly resistant or immune white pines will also minimize the occurrence of blister rust. Infected branches should be pruned as soon as possible. Small cankers on the main stem can be surgically excised by removing bark at least 1 inch (2.5 cm) beyond the canker margin. Trees with cankers extending more than half the trunk circumference should be removed.

GALL RUSTS

Fusiform Rust

Hosts   Fusiform rust is a serious disease of pines in the southern United States. Loblolly and slash pines are very susceptible, pitch and pond pines are moderately susceptible, longleaf pine is resistant, and shortleaf pine is immune to this disease. A large number of oak species serve as alternate hosts for this disease, including black, blackjack, English, live, post, red, scarlet, turkey, water, white and willow oaks.

Symptoms   Spindle-shaped swellings or galls appear on the branches and main stem (Fig. 11.5). Excessive branching is often associated with the galls. In late winter or early spring a yellow ooze forms over the gall and then orange-red blisters occur on the surface of the gall. Powdery yellow spores are produced in the blisters. This brightly colored stage persists for only a few weeks and then the bark returns to its normal dark color. Older galls become sunken as increased amounts of bark tissue are killed by the fungus. Branches and the main stem are often girdled and killed by the pathogen. Wind breakage frequently occurs at the galls on the trunk. On oak, orange pustules form on the underside

Fig. 11.5   Fusiform rust on loblolly pine seedlings. (Photo courtesy of George N. Agrios, University of Massachusetts, Amherst.)

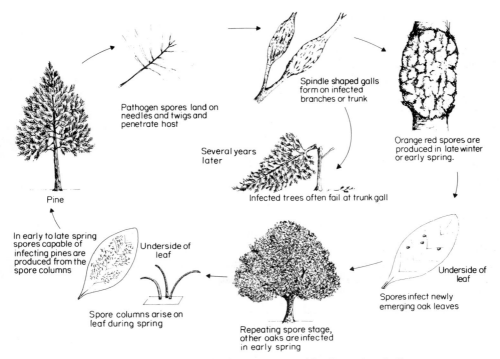

Pathogen spores land on needles and twigs and penetrate host

Spindle shaped galls form on infected branches or trunk

Orange red spores are produced in late winter or early spring.

Several years later

Infected trees often fail at trunk gall

Pine

In early to late spring spores capable of infecting pines are produced from the spore columns

Underside of leaf

Underside of leaf

Spores infect newly emerging oak leaves

Spore columns arise on leaf during spring

Repeating spore stage, other oaks are infected in early spring

Fig. 11.6 Disease cycle of fusiform rust caused by *Cronartium fusiforme*.

of the oak leaves. These are eventually replaced by brown, curled hairs that appear on the undersurface of oak leaves during the spring.

Disease Cycle Fusiform rust is caused by the fungus *Cronartium fusiforme* (Fig. 11.6). This fungus overwinters as mycelium in newly infected pine needles and in galls on living trees. In late winter or early spring fruiting bodies are formed below the bark and ooze a spore-containing exudate that allows fungal mating to occur. Soon after this spore stage, orange-red spores are formed, which are windblown to the expanding leaves of susceptible oaks. Pustules are formed on the underside of oak leaves about 2 weeks after infection. These pustules produce windblown repeating spores that infect other oak leaves. Later in the spring the production of repeating spores ceases and spore columns arise from the pustules. The spore columns produce windblown spores in early to late spring that can infect needles of susceptible pines by direct penetration. The fungus spreads through the needle, into the twig, and eventually into a branch or the main stem. Infected host tissues divide rapidly and swell to produce perennial galls.

Treatment Alternate host eradication of oaks is not feasible because suscepti-ble oaks are found growing naturally throughout the areas where this disease occurs, and many oaks are valuable shade trees. Avoidance of highly susceptible species of pine can minimize disease incidence. Planting stock should be examined carefully for galled seedlings, since infections in older trees started in seedlings. Nursery seedlings, however, can be protected against infection by several applications of fungicides during the spring. Infected branches should be pruned from the tree. Galls on the trunk can be surgically excised in the early stages before girdling. Matthews *et al.* (1976) found that excision of bark on fusiform rust galls back 1 inch (2.5 cm) into healthy bark was an effective method of eradicating the disease in high value trees.

Cedar–Apple Rust

Hosts Several junipers and cedars in the genus *Juniperus* are susceptible, including eastern and western red cedars and horizontal and savin junipers. Numerous members of the apple family serve as alternate hosts in this disease. Included are apple, chokecherry, crabapple, hawthorn, juneberry, mountain ash, pear, and quince. The disease on hawthorn and on quince is caused by another closely related species of the pathogen and is called cedar–hawthorn and cedar–quince rust, respectively, on these hosts.

Symptoms Brown galls form on the twigs and small branches of susceptible cedars and junipers. These galls are dimpled and range up to 2 inches (5 cm) across. In midspring small spikes protrude from each one of the dimples (Fig.

Fig. 11.7 (A) Cedar–apple rust gall on eastern red cedar in winter. (B) Cedar–apple rust gall in midspring. Note small bumps on gall surface from spore tendrils beginning to expand.

Fig. 11.8 Cedar–apple rust gall in late spring with orange spore tendrils is fully expanded during a wet period. (Photo courtesy of George N. Agrios, University of Massachusetts, Amherst.)

11.7). During wet weather these projections greatly expand into a brilliant, orange mass of jellylike tendrils, which can be easily seen from a distance (Fig. 11.8). These tendrils will shrivel during dry weather but can again expand during and shortly following wet periods several times in the spring. The mature galls dry out and harden during the summer. They no longer produce spores but may remain attached to the tree for several years. Yellow spots appear on the leaves of members of the apple family in late spring. On the upper surface of the leaves tiny pustules are formed in the spots while on the lower leaf surface small circular lesions bordered with ribbonlike strands are formed. These lesions on the lower leaf surface are easily visible with a hand lens and are referred to as "cluster cups."

Disease Cycle   Cedar–apple rust is caused by the fungus *Gymnosporangium juniperi-virginianae*. Similar diseases with hawthorn and quince as alternate hosts are caused by two other species of fungi in this genus. *G. globosum* and *G. clavipes*, respectively. These diseases have disease cycles similar to cedar–apple rust. All these fungi overwinter as thick-walled resting spores on the dimples of the gall (Fig. 11.9). During wet spring weather, when the spore tendrils are formed, these spores germinate and produce windblown spores that infect the leaves of susceptible trees in the apple family (Fig. 11.10). Infected areas on the leaves become yellow spots. Pustules are formed on the upper leaf surface and produce an ooze containing spores that is attractive to insects. As insects feed on the ooze, they spread spores from one pustule to those of another type and allow a type of mating to occur. Once this has occurred, cluster-cups form on the bottom of the leaf (Fig. 11.11). These cups will produce windblown spores during summer and early fall that are capable of infecting the foliage of susceptible cedars and junipers. The foliage of these trees is invaded by the pathogen and over a period of approximately 20 months a gall is formed. No repeating spore stage is formed in this disease. The gall is composed mostly of plant tissue that has greatly enlarged due to the influence of the pathogen. During the spring of the second season following infection, the galls mature and produce spore

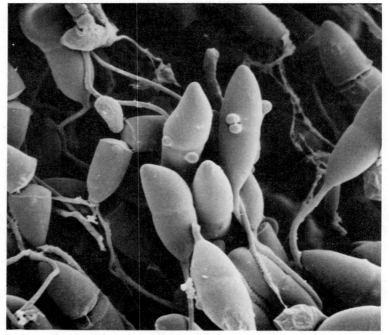

Fig. 11.9   Resting spores of *Gymnosporangium juniperi-virginianae* (× 750). (SEM courtesy of Merton F. Brown and H. G. Brotzman, University of Missouri, Columbia.)

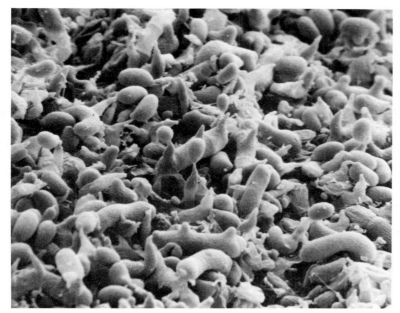

Fig. 11.10 Germination of resting spores of *G. juniperi-virginianae* and production of windblown spores that can infect apple trees (× 750). (SEM courtesy of Merton F. Brown and H. G. Brotzman, University of Missouri, Columbia.)

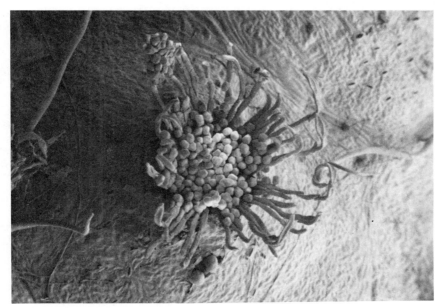

Fig. 11.11 Cluster-cup of *G. juniperi-virginianae* on bottom of an apple leaf (× 175). (SEM courtesy of Merton F. Brown and H. G. Brotzman, University of Missouri, Columbia.)

tendrils. The pathogen spends approximately 21 months on the juniper or cedar host and 3 months on leaves of trees in the apple family.

Treatment   This disease can be prevented by alternate host eradication of either the coniferous or hardwood host. However, such eradication is seldom necessary, except around a commercial fruit orchard, because little harm is done to either host. Infected cedars are considered attractive by some homeowners, especially during spring when the bright orange spore tendrils form. However, galls may be pruned from a healthy infected tree if their appearance is considered unattractive by the homeowner. Both coniferous and hardwood hosts can be protected from infection with fungicide applications during periods of spore production. Hardwoods should be sprayed just after bud break and at weekly intervals for 4 or 5 weeks depending upon rainfall frequency. Conifers should be sprayed three times at 2-week intervals beginning in mid-July.

LEAF RUSTS

Ash Leaf Rust

Hosts   Black, green, red, and white ashes are susceptible and species of salt marsh grass or cord grass serve as the alternate hosts for ash leaf rust.

Symptoms   In late spring yellow-orange spots occur on the upper surface of ash leaflets and also on petioles and shoots. About 2 weeks later, yellow-orange

Fig. 11.12   Ash leaf rust on upper and lower leaf surface of white ash. (Photo courtesy of Shade Tree Laboratories, University of Massachusetts, Amherst.)

Fig. 11.13 Distortion of petioles and shoots from infections of ash leaf rust. (Photo courtesy of Shade Tree Laboratories, University of Massachusetts, Amherst.)

cluster-cups appear below the spots on the under surface of leaflets and on the spots of the petioles and shoots (Fig. 11.12). If the infection is mild, with only a few scattered infections on the leaves, petioles or shoots, the tissues will remain alive through the summer. During heavy infections, however, the leaflets become curled and distorted from the many infections and the petioles and shoots can become girdled (Fig. 11.13). Leaf abscission begins in early summer and severely infected trees can be completely defoliated by midsummer. Many defoliated trees will refoliate, at least partially, by late summer. Yellow-orange blisters form on salt marsh grass during early summer. These blisters increase in number until late summer when the blisters turn black and remain in this condition during the winter on the dead marsh grass leaves.

Disease Cycle   Ash leaf rust is caused by the fungus *Puccinia peridermiospora* (Fig. 11.14). This fungus overwinters as thick-walled resting spores on the leaves of salt marsh grass. In midspring the resting spores produce windblown spores capable of infecting ash leaves, petioles, and shoots. The spores penetrate the ash tissues and form spots. Fruiting bodies are formed in these spots that ooze a spore-bearing exudate. Insects are attracted to this exudate and serve to spread the spores from one fruiting body to another, in a similar manner as in cedar–apple rust. Shortly after this spore stage is complete, open fruiting bodies called cluster-cups form on the underside of the leaf and on other infected tissues. These cups produce orange spores that are blown to salt marsh grass. These spores infect the marsh grass and cause blisters to form on the leaves. These blisters produce orange repeating spores, which are able to infect other marsh grass leaves. Repeating spores are replaced with black thick-walled overwintering spores in late summer.

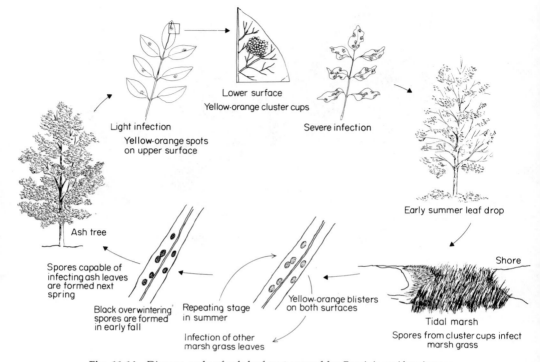

Fig. 11.14 Disease cycle of ash leaf rust caused by *Puccinia peridermiospora*.

Treatment   Ash leaf rust usually does not cause sufficient injury to ash trees or to salt marsh grass to recommend either prevention or control measures. If the disease causes severe defoliation for successive years, however, the foliage should be protected to prevent decline of the tree. Ash trees may be protected, at least partially, by fungicide applications made at weekly intervals while the foliage is developing in mid to late spring. A homeowner may also wish to protect a high-value ash for aesthetic purposes. Although ash rust is found over much of the eastern United States and Canada, it is most severe near the sea coast where the alternate host, salt marsh grass, is in great abundance. In these areas ash leaf rust is a perennial problem that can reach epidemic proportions during successive wet seasons. Under these conditions planting ash trees should be avoided.

Red Pine Needle Rust

Hosts   A large number of 2- and 3-needled pines are susceptible, including Coulter, jack, lodgepole, pitch, ponderosa, and table-mountain pines. Closely related needle rusts also occur on other pines in this group, such as loblolly,

Fig. 11.15   Red pine needle rust on red pine. (Photo courtesy of USDA Forest Service.)

Scots, shortleaf, and Virginia pines. Goldenrod and aster are the alternate hosts for this disease.

Symptoms   White pustules form on needles of susceptible pines in the spring (Fig. 11.15). These pustules erupt through the needle surface and extend above the needle. Light infections may go unnoticed, but heavy infections cause the entire tree to have a yellow-green color. Needle rust symptoms on pine are sometimes confused with scale insects. Rust pustules, however, are raised above the needle and cannot be scraped off easily with a fingernail, while scale insects are closely appressed to the needle and come off easily when scraped with a fingernail. Orange-red pustules form on goldenrod and aster in early summer. These pustules remain through the summer and are replaced by flattened red-brown pustules in the fall.

Disease Cycle   Red pine needle rust is caused by the fungus *Coleosporium asterum* (Fig. 11.16); related pine needle rusts are caused by other *Coleosporium* species. This fungus overwinters as mycelium in infected pine needles. In the spring tiny pustules ooze a spore-bearing exudate (Fig. 11.17). This stage allows the mating of different spore types as described earlier in ash rust and cedar–apple rust. Raised, white pustules that contain yellow spores form on the needles (Fig. 11.18). The spores are blown to aster or goldenrod leaves in late spring or early summer. Pustules are formed on infected leaves and produce repeating spores during the summer that reinfect other aster or goldenrod leaves

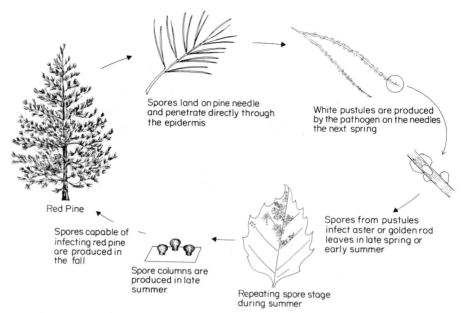

Spores land on pine needle and penetrate directly through the epidermis

White pustules are produced by the pathogen on the needles the next spring

Red Pine

Spores capable of infecting red pine are produced in the fall

Spore columns are produced in late summer

Repeating spore stage during summer

Spores from pustules infect aster or golden rod leaves in late spring or early summer

Fig. 11.16 Disease cycle of red pine needle rust caused by *Coleosporium asterum.*

Fig. 11.17 Pustule of *C. asterum* on pine needle (× 500). (SEM courtesy of Merton F. Brown and H. G. Brotzman, University of Missouri, Columbia.)

Fig. 11.18 (A) Raised pustule of *C. asterum* containing spores that will infect aster and goldenrod leaves (× 125). (B) Close-up of spores (× 500). (SEM courtesy of Merton F. Brown and H. G. Brotzman, University of Missouri, Columbia.)

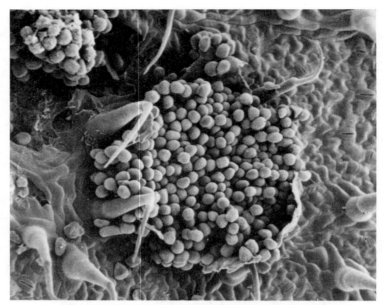

Fig. 11.19 Open pustule of *C. asterum* containing repeating spores (× 225). (SEM courtesy of Merton F. Brown and H. G. Brotzman, University of Missouri, Columbia.)

Fig. 11.20 Spore columns of *C. asterum* producing windblown spores that will infect susceptible pines (× 225). (SEM courtesy of Merton F. Brown and H. G. Brotzman, University of Missouri, Columbia.)

TABLE 11.1

Some Additional Rust Diseases of Trees[a]

Disease	Host	Alternate host	Pathogen	Symptoms
Commandra blister rust	2- or 3-needled pines	Bastard toad flax (*Commandra* spp.)	*Cronartium commandrae*	Spindle-shaped swellings on branches and trunks
Sweetfern blister rust	2- or 3-needled pines	Sweetfern (*Comptonia perigninia*)	*Cronartium comptoniae*	Orange pustules on stem or on twig galls; most damaging by its girdling action; after a tree reaches basal diameter of about 3 inches (7.5 cm), it is relatively safe
Eastern gall rust (Pine–oak rust)	2- or 3-needled pines	Oak	*Cronartium quercum*	Knotty galls on stems or trunks
Western gall rust (Pine–pine rust)	2- or 3-needled pines	None	*Peridermium harknessii*	Knotty galls on stems or trunks
Aspen rust	Poplars	Numerous conifers	*Melampsora medusae*	Yellow pustules on leaves that cause defoliation when they cover more than half of the leaf surface
Fir-bloom rust	Balsam fir	Chickweek (*Cerastii* spp.)	*Melampsorella cerastii*	Witches' broom with dwarfed and yellowish needles; two rows of orange-yellow pustules on lower needle; needles shrivel and drop
Fir-huckleberry rust	Balsam fir	Blueberry (*Vaccinium* spp.)	*Pucciniastrum geoppertianum*	White-yellow pustules on lower surface of current needles
Fir-fern rust	Balsam fir	Sensitive fern (*Onoclea sensibilis*)	*Uredinopsis mirabilis*	White, cylindric pustules on lower surface of current needles; needles turn brown
Spruce needle rust	Spruce (white and blue)	Labrador tea (*Ledum* spp.)	*Chrysomyxa* spp.	Whitish blisters on lower surface of 1-year needles; needles turn yellow and may drop prematurely

[a] From Blanchard, Tattar, and MacHardy (1974).

(Fig. 11.19). In late summer red-brown spore columns replace the repeating spores. In the fall the spore columns produce windblown spores that infect needles of susceptible pines (Fig. 11.20).

Treatment   Red pine needle rust usually causes little harm to either host and control measures are normally not recommended. Severe infections occasionally occur when both hosts are in close proximity. Severe infections can cause loss of aesthetic value of pines and are a cause of concern for the homeowner. In these cases, eradication of asters and goldenrods from around the pines will often decrease severity. Pine needles can also be protected from infection by fungicide applications in late summer and early fall.

OTHER RUST DISEASES

A large number of rust diseases occur on shade trees. Only a few examples from the most common rust diseases of trees were presented in this chapter. A list of some additional rust diseases is given in Table 11.1.

LITERATURE CITED

Blanchard, R. O., T. A. Tattar, and W. E. MacHardy. (1974). Identification and control of Christmas tree diseases in New Hampshire. *Univ. N.H., Coop. Ext. Serv., Publ.* **23**.

Matthews, F. R., H. R. Powers, and B. G. Arnold. (1976). Eradication of fusiform rust on loblolly pine rootstock—comparison of sodium arsenite and bark excision. *Tree Planters' Notes* **2**, 9–10.

SUGGESTED REFERENCES

Anderson, G. W. (1970). Sweetfern rust on hard pines. *U.S., For. Serv., For. Pest Leafl.* **79**, 1–6.

Blanchard, R. O. (1974). Ash leaf rust. *For. Notes* **118**, 27–28.

Czabator, F. J. (1971). Fusiform rust of southern pines—a critical review. *U.S., For. Serv., Res. Pap.* **SO-65**, 1–39.

Davison, A. D. (1969). Needle rust of grand fir. *Proc. Int. Shade Tree Conf.* **45**, 257–258.

Mielke, J. L., R. G. Krebill, and H. R. Powers. (1968). Commandra blister rust of hard pines. *U.S., For. Serv., For. Pest Leafl.* **62**, 1–8.

Nicholls, T. H., and R. L. Anderson. (1976). How to identify and control pine needle rust disease. *U.S. For. Serv. Leafl.* 1–8.

12

Wilt Diseases

INTRODUCTION

Wilt diseases can kill a large, healthy tree during a single growing season. Diseases that affect the vascular system of a plant are called wilt diseases. The attack on the vascular tissue by the fungus causes moisture stress that eventually leads to wilting. Since the vascular system performs the vital function of transport within the plant, diseases of the vascular system can cause rapid killing of large branches and even entire trees. The impact of wilt diseases of shade trees has been so great that many homeowners think of wilt diseases of trees whenever the subject of plant diseases is mentioned.

TYPICAL WILT DISEASE CYCLE

Wilt diseases can be separated into two major groups: (1) those that begin in the branches as a result of leaf or bark feeding by pathogen-infested insects; and (2) those that begin in the roots as a result of wounding or possibly direct root penetration by the fungus. Infection can occur in both these groups by root grafts. However, once inside the host tissues the pathogen remains almost exclusively in the vascular system as long as the host is living. Wilt disease fungi are disseminated through the vessels in the outer rings of xylem where most of the transport of sap occurs and causes increased loss of functional xylem. When the fungi have invaded most of the vessels around a branch, it quickly wilts and dies. Wilt fungi that invade twigs cause wilting in small and progressively larger branches, often in a short period. Wilt fungi that invade roots often cause symptoms similar to wilts starting in twigs, but in some cases large sections or the entire tree may suddenly wilt as the pathogen completely kills the vascular tissue in the main stem. The pathogen in the roots must pass through the trunk and main branches before it can cause wilting in small branches. There is a large amount of variability in the rate of symptom development in wilt diseases. Host susceptibility varies considerably in most wilt diseases both within and between species.

Wilt disease fungi are facultative parasites that can live on the host after it has died. The dead host, therefore, remains a source of inoculum for insect vectors or a center for the spread of soil pathogens for several years. Wilt fungi in the trunk and branches can be carried to new hosts by insect vectors. These fungi can also invade healthy roots of adjacent trees of the same species through root grafts.

TYPICAL CONTROL OF WILT DISEASES

Wilt diseases are controlled primarily by protection of susceptible trees against insect vectors, by eradication of infected hosts to minimize disease spread, and by prevention of infection by root grafts. In many cases control programs have to be coordinated by various governmental agencies to be effective in protecting a large population of susceptible trees. Early diagnosis through surveillance and prompt removal of infected trees are critical to the success of any wilt disease control program. Since wilt fungi can often invade healthy trees through root grafts, these must also be severed whenever the roots of an infected tree are in possible contact with another of the same species.

Therapeutic treatments of trees infected with wilt diseases have been attempted by researchers and arborists for many years. The success of any of these treatments or "cures" has been minimal. The recent development of systemic fungicides, however, has given new hope to the possibility of successful wilt disease therapy. Systemic fungicides move through the vascular system and are most effective when the tree is completely healthy. In this condition the entire vascular system is functional and the material moves throughout the outer trunk and branches. When a branch is infected with a wilt pathogen it causes dysfunction of the vascular system and the systemic chemical cannot reach the areas of the tree where it is most needed. The chemical does prevent, at least temporarily, the further invasion of the pathogen into healthy vascular tissue. Systemic chemicals, therefore, can only be effective in the early stages of a wilt disease when most of the vascular system is still healthy. These chemicals are of little value in the advanced stages when most of the vascular system is dead. Applications of systemic chemicals have yielded some positive results but are still in the experimental stage. They should not be substituted for standard control practices of insect vector control and sanitation, but added to these practices as an additional tool to control wilt diseases.

WILT DISEASES OF TREES

Dutch Elm Disease

Hosts   North American species of elm, such as American, slippery, and winged elms, are highly susceptible, while European elms such as Dutch, English, and Scotch elms are moderately resistant, and Asiatic elms such as Chinese and Siberian are highly resistant. Resistant hybrids have been developed that incorporate the genetic resistance of the smaller Asiatic elms with the larger size and more desirable shade tree form of the American elm or European elms. Most of these hybrid elms have been produced in the Netherlands, such as the Christine Buisman, Commelin, and Groeneveld elms, but the University of Wisconsin has released the Autumn Gold elm and the Shade Trees and Ornamental Plants Laboratory of the U.S. Department of Agriculture has recently released the Urban elm. In addition, some promising new hybrids from the Netherlands are currently in postentry quarantine and should be available in the next few years.

Symptoms   The earliest symptoms on the tree are yellowing and/or wilting of the leaves on a single branch usually in the upper crown. The yellow or wilted leaves quickly turn brown and die. This symptom is known as a "flag." The symptoms rapidly spread, usually one branch at a time, to progressively larger branches and eventually the entire tree (Fig. 12.1). If initial symptoms begin in late spring the tree usually will be dead by the end of the summer, although large trees sometimes take several years to die. If symptoms begin in midsummer disease progression will often be limited to only a few branches. The next season, however, symptom progression often recurs in late spring and the trees are killed by late summer. If initial symptoms begin in late summer the disease will usually not progress any further during that season. Such a tree often remains free of symptoms the next season.

There exists a great deal of variability of symptom development in the Dutch elm disease. In addition to the size of the tree and the time of year that infection begins, the susceptibility of the host, the virulence of the pathogen, and the site of initial infection also strongly influence disease development. The symptom progressions previously described are for highly susceptible North American elm species that have been infected in the twigs. In species with some resistance to the disease, such as the European elms, symptom development is much slower and sometimes the trees recover. Infection can also occur in large branches, the trunk, or through the roots. In these cases symptoms may develop as a general decline, with yellowing followed by wilting in the entire tree. These symptoms are sometimes confused with similar ones produced by elm phloem necrosis.

Symptomatic branches contain brown discoloration or "streak" in the outer xylem (Fig. 12.2). This type of discoloration is a common symptom of vascular

Fig. 12.1 Dutch elm disease in American elm. Note defoliation and wilting on the left half of the tree and normal appearance on most of the right half. Inset, close-up of wilted and curled leaves. (Photo courtesy of Shade Tree Laboratories, University of Massachusetts, Amherst.)

Fig. 12.2 Discoloration in outer xylem of an elm twig taken from a tree with Dutch elm disease. (Photo courtesy of Shade Tree Laboratories, University of Massachusetts, Amherst.)

wilt diseases caused by fungi. Small twigs less than 1 inch (2.5 cm) in diameter should be removed from affected branches. If the outer wood is shaved with a knife a band or scattered lines of discoloration are revealed in infected twigs. Discoloration can also be found in larger branches, the trunk, and sometimes the roots of severely infected trees. In dead trees or trees infected more than one season, egg galleries of bark beetles can be found beneath recently killed bark (Fig. 12.3).

Disease Cycle   Dutch elm disease is caused by the fungus *Ceratocystis ulmi* (Fig. 12.4). This fungus overwinters in infected and recently killed trees, in stumps, and in recently cut brush and logs. The fungus is carried from infected wood to healthy trees by elm bark beetles. Two species of elm bark beetles are important vectors of this pathogen, the European and the native elm bark beetle. Both these insects lay their eggs under the bark primarily in stressed, dead, or dying elms. In the late fall the eggs hatch and produce larvae that tunnel beneath the bark. In early spring the larvae form the pupal "resting stage" and emerge as adults in mid to late spring. In infected trees the fungus produces balls of sticky spores in the beetle galleries (Fig. 12.5). The bodies of the adult

Fig. 12.3   Egg galleries of the European elm bark beetle beneath the bark of an American elm. (Photo courtesy of Shade Tree Laboratories, University of Massachusetts, Amherst.)

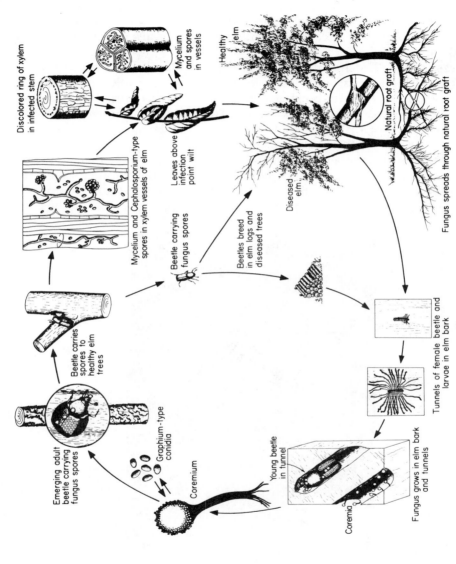

Fig. 12.4 Disease cycle of Dutch elm disease caused by *Ceratocystis ulmi*. [Drawing courtesy of George N. Agrios. (1969). Plant pathology. Academic Press, New York.]

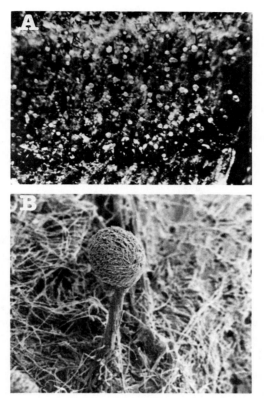

Fig. 12.5 (A) Balls of sticky spores (coremia) of *C. ulmi*. (B) Close-up of a coremium (× 300). [(A) Photo courtesy of Shade Tree Laboratories, University of Massachusetts, Amherst, and (B) SEM courtesy of Merton Brown and H. G. Brotzman, University of Missouri, Columbia.]

elm bark beetles are contaminated with these spores when they emerge. The European beetle feeds primarily on the bark in small twig crotches in the upper crown (Fig. 12.6). The native elm beetle feeds primarily on larger branches and sometimes on the main trunk. The fungus on the insect's body is introduced into the elm tree through the feeding wounds. The fungus enters the vascular system and progressively grows through the outer xylem vessels. The fungus produces spores (cephalosporium-type) in the vessels that are thought to increase its rate of spread within the tree. As the fungus invades more vascular tissue, twigs, branches, and the entire tree wilt and die. Once a tree is dead the fungus will invade wood all over the tree, including the roots. The dying or dead tree is attractive to the female elm bark beetle looking for a place to lay eggs for the next brood. Two or three broods may occur in a single year. The overwintering brood, however, is most important for the spread of the pathogen since the trees

Fig. 12.6 European elm bark beetle feeding in a twig crotch of American elm. (Photo courtesy of Shade Tree Laboratories, University of Massachusetts, Amherst.)

are most susceptible when it emerges in the spring. The pathogen may also invade a healthy tree through root grafts from an adjacent infected tree.

Treatment The incidence of Dutch elm disease in a community can be minimized by an organized program of insect vector control and prompt removal of infected hosts. A community effort is stressed because these control measures are most effective when applied to all the susceptible trees in an area. Elm bark beetles are most effectively controlled by dormant spraying in the spring. Early detection of infected trees is essential to prevent disease spread and can be achieved through a program of surveillance for initial symptoms. Severely infected trees should be removed as soon as possible and the wood buried or burned before the next spring. Any possible root grafts (Fig. 12.7) should be severed if a healthy elm tree is within 50 feet (16 m) of an infected one. This can be done mechanically by digging a trench 3 ft (1 m) deep or chemically by infection of a soil fumigant 18 inches (45 cm) deep, in a line between the trees.

An infected tree can often be saved in the early stages of the disease by removing all infected branches. This therapeutic procedure is most effective when less than 5% of the tree is wilted. All wilted branches must be removed as well as any branches with discolored bands or streaks in the outer xylem. The pruning cuts should extend at least 1 foot (30 cm) beyond the last visible point of

Fig. 12.7 Root graft (bark removed) between two adjacent American elms. (Photo courtesy of Shade Tree Laboratories, University of Massachusetts, Amherst.)

discoloration. Recently, systemic chemicals that can stop the growth of the pathogen in the tree, at least temporarily, have been developed. These materials injected into the trunk of a mildly infected tree (less than 10% wilt) can sometimes stop symptom progression of Dutch elm disease. Systemic chemicals are most effective when used in combination with pruning of infected branches. Kondo (1972) has found injection through roots also to be an effective way to add these materials to a tree. Although these materials appear promising, they are still experimental. Treatments with systemic chemicals are most effective when used to prevent infection in healthy trees but for continued protection treatments must be reapplied each year.

Resistant elm species and resistant elm hybrids can be substituted for highly susceptible native elm species. The vase form of the American elm, however, is not found in the resistant species or hybrids. The Zelkova, a close relative of the elm, is vase-shaped and resistant to the Dutch elm disease. It is sometimes recommended as a substitute for the American elm; however, this species is sometimes attacked by numerous target cankers.

Oak Wilt

Hosts All oak species are susceptible but species in the red oak group, including black, pin, red, scarlet, and shingle oaks, are very susceptible, while species in the white oak group, including burr, chestnut, post, swamp white, and white oaks, are fairly resistant to the disease although susceptibility varies within the group.

Symptoms Symptom development differs between the red and white oak groups. In the red oak group, the leaves turn a dull, water-soaked green, then yellow or brown, beginning at the tip and proceeding toward the petiole. Symptomatic leaves are shed in all stages at this time. Symptom development begins in mid to late spring, usually starting at the top of the tree, but quickly spreads throughout the tree (Fig. 12.8). Symptom development may occur so rapidly as

Fig. 12.8 Defoliation of a red oak from oak wilt. (Photo courtesy of USDA Forest Service.)

to appear over the whole tree at once. Sprouts or suckers may appear on the trunk and large branches after the tree has lost most of its leaves. An infected tree may die in as little as a few weeks or over the summer. In the white oak group, similar foliar symptoms occur but only one or a few branches are affected each season. The symptomatic branches are killed each year and the tree declines over several years and eventually dies, but in some cases an infected tree may recover. Brown-black discoloration can sometimes be found in the outer sapwood by peeling the bark on twigs or branches that exhibit foliar symptoms. Brown-black discoloration is often found in the outer sapwood of twigs in the white oak group but is not a common symptom in the red oak group. Raising and cracking of the bark occur in dead trees. Removal of the bark in these areas will usually reveal oval mats or pads of the fungus mycelium, which produce a fruitlike odor that is attractive to insects.

Disease Cycle   Oak wilt is caused by the fungus *Ceratocystis fagacearum* (Fig. 12.9). The fungus overwinters as mycelium in infected trees and as fungus pads on dead trees. In the spring the pads produce large numbers of spores that are carried to healthy susceptible trees by insects, which feed on the pads. Sap- and

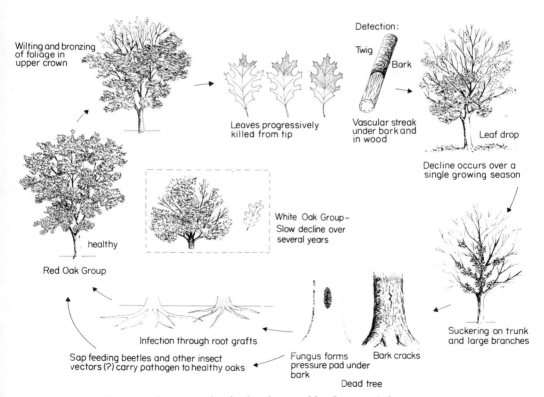

Fig. 12.9   Disease cycle of oak wilt caused by *Ceratocystis fagacearum*.

bark-feeding beetles contaminated with *C. fagacearum* spores introduce the fungus into healthy trees through wounds caused by feeding. The pathogen spreads rapidly within the xylem vessels and causes foliar symptoms to occur. After the tree is killed the fungus grows throughout the outer wood and also invades the root system. Healthy susceptible oaks growing close to an infected oak can become infected through root grafts. Mycelial mats usually form under the bark a few months after the tree is killed and, through outward pressure against the bark, force it to crack. Insects enter through the bark cracks and feed on the pads.

Treatment   Oak wilt is often controlled most effectively, like Dutch elm disease, through a community program of early detection and prompt removal of dying and dead trees. Spread of the pathogen by insects can be minimized by avoiding all wounds, including pruning, from midwinter to early summer. Oaks are susceptible to infection during this period and sap-feeding insects that transmit the spores of the pathogen are attracted to fresh wounds. If wounding occurs by accident or cannot be avoided the wound should be immediately covered with a wound dressing to prevent insect feeding. Species in the red oak group rarely survive more than a few months after infection and should be removed as soon as possible. Species in the white oak group, however, often live several years after infection and in some cases appear to have recovered. If an infected white oak is a valuable shade tree it need not be removed unless its aesthetic value is lost or it becomes a hazard tree. Local spread of the pathogen through root grafts must also be prevented. A narrow trench approximately 3 feet (1 m) deep should be dug between a healthy and an infected oak, of the same group, that are 35 feet (10 m) or less apart. An alternate method is to kill a small band of roots between the trees with a soil fumigant. However, there are no therapeutic measures recommended for a tree once it has become infected.

Verticillium Wilt

Hosts   A large number of woody plants are susceptible to Verticillium wilt. Some of the common woody hosts include ash, black locust, catalpa, dogwood, elm, magnolia, maple, red bud, Russian olive, sumac, tulip tree, viburnum, yellow poplar, and yellow wood. Verticillium wilt is also a common disease of many crop plants such as cotton, eggplant, potato, and tomato. Resistant species of woody plants include beech, birch, crabapple, European mountain ash, ginkgo, hackberry, hawthorn, holly, honey locust, Katsura, mulberry, oak, pecan, pyracantha, sweet gum, sycamore, willow, and zelkova. Conifers are immune to the disease.

Symptoms   The most common initial symptom is the sudden wilting of one or more branches on a tree. Prior to wilting some branches may produce small leaves that are often cup-shaped. Within a few weeks an entire side of the tree and in some cases the entire tree may wilt (Fig. 12.10). Symptoms often progress rapidly in small trees, which are usually killed in two or three seasons following

Fig. 12.10 Verticillium wilt on sugar maple. Early wilt symptoms (A). Branch dieback late in the growing season (B).

infection. Large trees often survive many years after initial infection and in some cases a tree may only lose a few branches each season. The xylem in symptomatic branches usually contains abundant discoloration. The color will vary with the species attacked, for example, infected maple wood will be olive green to green-black; black locust wood will be brown to black; and elm wood will be brown. Occasionally small branches will completely lack discoloration even in a completely wilted tree but abundant discoloration can be found in the trunk and main branches. Discoloration can also be followed into the root system. Long diffuse cankers occasionally form on the bark around symptomatic branches. Sprouts also may occur below and opposite wilted branches and at the base of the tree (Fig. 12.11).

Disease Cycle Verticillium wilt is caused by the fungus *Verticillium alboatrum* or *V. dahliae* (Fig. 12.12). These fungi overwinter in infected hosts and in the soil as saprophytes. The pathogen can enter the tree through wounds in the roots or near the soil line, through root grafts and possibly through direct penetration of a healthy root (Fig. 12.13). Once in the tree the fungus moves through the vascular system up the trunk and into the branches (Fig. 12.14). Although

Fig 12.11 Verticillium wilt on American elm. Upper branches have been defoliated but abundant sprouts have occurred on the main branches. (Photo courtesy of Shade Tree Laboratories, University of Massachusetts, Amherst.)

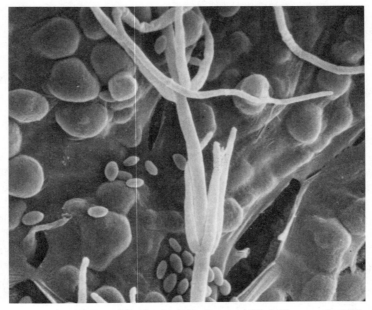

Fig. 12.12 Spore-bearing hypha (conidiophore) of *Verticillium* sp. (× 200). (SEM courtesy of Merton Brown and H. G. Brotzman, University of Missouri, Columbia.)

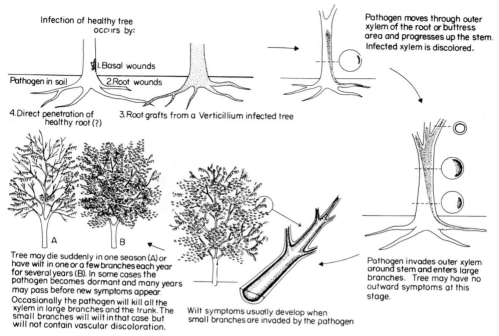

Infection of healthy tree occurs by:

1. Basal wounds

Pathogen in soil

2. Root wounds

4. Direct penetration of healthy root (?)

3. Root grafts from a Verticillium infected tree

Pathogen moves through outer xylem of the root or buttress area and progresses up the stem. Infected xylem is discolored.

Tree may die suddenly in one season (A) or have wilt in one or a few branches each year for several years (B). In some cases the pathogen becomes dormant and many years may pass before new symptoms appear. Occasionally the pathogen will kill all the xylem in large branches and the trunk. The small branches will wilt in that case but will not contain vascular discoloration.

Wilt symptoms usually develop when small branches are invaded by the pathogen

Pathogen invades outer xylem around stem and enters large branches. Tree may have no outward symptoms at this stage.

Fig. 12.13 Disease cycle of Verticillium wilt caused by *Verticillium albo-atrum* or *V. dahliae*.

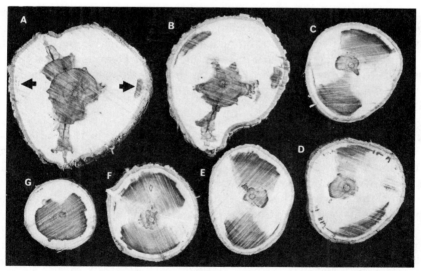

Fig. 12.14 Cross sections of the trunk of a symptomless Norway maple tree that was inoculated with *V. albo-atrum* (see arrows) 10 cm aboveground. (A) 10 cm; (B) 50 cm; (C) 1 m; (D) 1.5 m; (E) 2 m; (F) 3 m; (G) 4 m aboveground. Peripheral discoloration that expands from (A) to (G) is associated with *V. albo-atrum*. Central discoloration is associated with a basal wound (A)–(C) and branch stubs (D)–(G).

the vascular system of the trunk and large branches may contain abundant discoloration the infected tree may remain symptomless. Continued invasion by the pathogen results in symptom development and eventual death of the host.

Treatment Verticillium wilt can be avoided in areas where it has been a problem by selecting species known to be resistant or immune. When a tree is infected with Verticillium wilt any possible root grafts between adjacent healthy trees should be cut as described earlier for Dutch elm disease and oak wilt. When a tree has died from Verticillium wilt as much of the root system as possible should be removed and the soil replaced or fumigated before replanting any susceptible species. *Verticillium* spp. are fungi that can live in the soil for many years after the tree has been killed.

There is no cure for Verticillium wilt; however, symptom progression can sometimes be delayed for many years by increasing the vigor of an infected tree through fertilization. Watering during dry periods in summer and fall is also recommended. Dead branches should also be pruned. This does not remove the fungus from the tree but helps to maintain its aesthetic value.

Mimosa Wilt

Hosts The silk tree or "mimosa" is the only woody plant susceptible to this disease. Mimosa varieties Charlotte and Tryon are resistant.

Symptoms Initial symptoms are wilting of leaves in one or two branches, usually in the upper crown. Wilted leaves quickly turn yellow, die, and are shed. Symptoms progress from one branch at a time until all the branches are killed, often by the end of the growing season or a year following initial symptom development. Symptomatic branches contain brown discoloration in streaks or in a ring just under the bark in the outer wood. Discoloration is most pronounced in the buttress and roots during the early stages of the disease. The bark may split and ooze may flow from the trunk during the final stages of the disease.

Disease Cycle Mimosa wilt is caused by the fungus *Fusarium oxysporum* form *perniciosum*. The fungus overwinters in infected hosts or in the soil as a saprophyte. The pathogen enters a healthy mimosa by wounds in the roots or at the base of the tree. Movement is through the xylem vessels up the tree, similar to Verticillium wilt. Branches wilt as a result of the progressive loss of functional vessels and the tree eventually dies. This fungus can persist in the dead root system for many years after the tree is killed.

Treatment The use of resistant varieties of mimosa will decrease incidence in new plantings. Once a tree has been killed by mimosa wilt another mimosa should not be planted in the same area, unless the root system of the dead tree is first removed and the soil fumigated prior to planting. There is no control for mimosa wilt once a tree has become infected and no therapeutic measures are known to affect symptom progression.

LITERATURE CITED

Kondo, E. S. (1972). A method for introducing water-soluble chemicals into mature elms. *Can. For. Serv. Inf. Rep.* **O-X-171**, 1–11.

SUGGESTED REFERENCES

Cannon, W. N., and D. P. Worley. (1976). Dutch elm disease control: Performance and costs. *U.S., For. Serv., Res. Pap.* **NE-345**, 1–7.

Fowler, M. E. (1958). Oak wilt. *U.S., For. Serv., For. Pest Leafl.* **29**, 1–7.

Gregory, G. F., and T. W. Jones. (1975). An improved apparatus for pressure-injecting fluid into trees. *U.S., For. Serv., Res. Note NE* **NE-214**, 1–6.

Hanisch, M. A., H. D. Brown, and E. A. Brown (eds.). (1983). Dutch elm disease management guide. *USDA Forest Serv., Bull.* **1**, 1–23.

Himelick, E. B. (1968). Verticillium wilt. *Proc. Int. Shade Tree Conf.* **44**, 256–262.

Himelick, E. B. (1969). Tree and shrub hosts of Verticillium albo-atrum. *Ill. Nat. Hist. Surv., Biol. Notes* **66**, 1–8.

Himelick, E. B. (1972). High pressure injection of chemicals into trees. *Arborist's News* **37**, 97–103.

Himelick, E. B., and D. Neely. (1965). Prevention of root graft transmission of Dutch elm disease. *Arborist's News* **30**, 9–13.

Holmes, F. W. (1976). The American elm fights back. *Horticulture* **54**, 72–78.

Schreiber, L. R., and J. W. Peacock. (1974). Dutch elm disease and its control. *U.S., Dep. Agric., Agric. Res. Serv., Bull.* **193**, 1–15.

Sinclair, W. A., and R. J. Campana (eds.). (1978). Dutch elm disease—Perspective after 60 years. Northeast Regional Research Publication, Cornell University, Search Vol. 8, No.5. 52p.

Van Alfen, N. K., and G. S. Walton. (1974). An evaluation of the Lowden formulation containing nystatin for Dutch elm disease control. *Plant Dis. Rep.* **58**, 924–926.

Wilson, C. L. (1975). The long battle against Dutch elm disease. *J. Arbori.* **1**, 107–112.

Wilson, C. L. (1976). Recent advances and setbacks in Dutch elm disease research. *J. Arbori.* **2**, 136–139.

13

Wound Diseases— Discoloration and Decay in Living Trees

INTRODUCTION

A wound is the first step in a complex series of events that often leads to discoloration and decay of wood in living trees. A wound is a break in the bark where the wood underneath is exposed. While some wounds result in very little discoloration, other wounds result in extensive discoloration and decay—but in all cases this process starts with a wound. Discoloration and decay have been studied extensively in living trees but most of the emphasis until recently has been on decay, which is the last step in the process. Recent research on wounding has indicated that the type and severity of the wound and the wound response of the host are the best indicators of the potential amount of discoloration and decay that will occur in a tree. This chapter will discuss the various types of wounds that occur commonly on shade trees and will also discuss means to treat wounds to minimize discoloration and decay.

The arborist is concerned with preventing decay in wounded trees and determining physical stability in trees that have internal decay. Detection of hazard trees is a critical and complex activity, in that it includes other categories of defect, in addition to internal decay (see Chapter 26).

The arborist often has to rely upon the presence of a fruiting structure of a decay fungus (mushroom or conk), exposed decayed wood on an old wound, or a cavity, to detect decay in living trees. Decayed wound faces and cavities often represent advanced internal decay far beyond that exposed to the observer. Fruiting bodies of decay fungi are unreliable indicators of decay because they often do not appear until decay is well advanced if at all. Examination of wounds as predictors of defect has proven to be an accurate and relatively easy procedure.

WOUNDS

Shade trees are constantly being wounded, mainly due to their proximity to the activities of people. Some common sources of wounding are automobiles, bicycles, birds, cats, dogs, fire, lawnmowers, people, and snowplows. Most wounds are small and close quickly; however, some wounds are severe and require attention to close properly. Since all trees lose many branches during their lifetime, either by breakage or by pruning, trees are exposed to many branch wounds. Branch wounds are often overlooked. A broken branch stub is a serious wound that, if untreated, often results in decay.

EVENTS AFTER WOUNDING

Host Response

Trees respond to wounds as soon as they occur. Trees respond on several levels: first, a rapid electrical and chemical response; later, several structural responses. These various responses are designed for two general purposes: (1) to block the invasion of microorganisms after wounding, (2) to confine or compartmentalize the wound. The success of the host response depends to a great extent on the severity of the wound, and on the genetic potential of the tree to exhibit a strong wound response. In general, small wounds close quickly while larger wounds are more severe, close slowly, and often result in discoloration and decay, but exterior wound size is not the only indicator of severity. Some types of small wounds can result in considerable disruption of woody tissues and are more severe than large superficial wounds.

The anatomical response of the tree to a wound is termed *compartmentalization*. The tree confines the invasion of microorganisms to the injured tissues by placing several barriers around them. By "walling off" the area around the wound, the tree places the wound in a sealed compartment and prevents or minimizes any microbial invasion in the rest of the tree. It is, however, difficult to think of the response of the host to a wound without also considering the active role of the microorganisms that invade the wound. The ability of the host to compartmentalize the wound often depends on how quickly barriers can be placed in front of the invading microorganisms (Fig. 13.1).

The tree, in some ways, is constructed like a large ship. The interiors of both a tree and a ship are composed of compartments that can be shut in case of emergency to prevent the loss of the entire structure. The tree, in contrast to a ship, also produces a completely new layer of bark and wood each year by the action of the cambium. The cambium forms the strongest barrier and prevents any microorganisms from invading tissues that are formed after wounding (Fig. 13.2). The cambium responds to wounding by changing the type of cells it produces to create a barrier zone. Inside the tree the rays form walls of smaller

Fig. 13.1 (A) Well-closed branch wounds with little discoloration and no decay behind the wound. Note callus layer of clear wood over wound. (B) Poorly closed branch wound. Note extensive column of discoloration and decay behind the wound. (Photos courtesy of Alex L. Shigo, USDA, Forest Service, Durham, New Hampshire.)

interior compartments that block movement of pathogens in the lateral direction. The floor and ceiling of the compartments are formed soon after wounding. Hollow vessel cells in the wood that normally carry sap to the leaves become plugged above and below the wound. These plugged cells inhibit the invasion of the pathogen. Chemicals inhibitory to the pathogens are produced by living cells in the wood in advance of the pathogen. The barriers created in the vessels and along the rays are often only temporary barriers and eventually fail. However, the host continues to respond in the vessels and in the rays in advance of the invading microorganisms, constantly slowing their movement and limiting their spread within the tree (Fig. 13.3). The host responses confine the invading microorganisms to as small a part of the tree as possible. After many years the barrier zones at the rays and in the vessels often break down and a hollow forms (Fig. 13.4), but the barrier zone formed at the cambium remains effective, and tissues formed after wounding are sound. Compartmentalization allows most

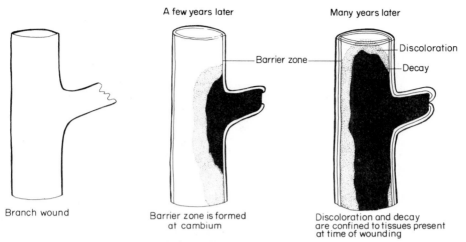

A few years later   Many years later

Discoloration

Barrier zone

Decay

Branch wound   Barrier zone is formed   Discoloration and decay
at cambium   are confined to tissues present
at time of wounding

Fig. 13.2 Compartmentalization of wounds.

Fig. 13.3 Limitation of microorganism invasion within a defect column. Although decay fungi have created a small hollow in the center of the tree the extent of microbial invasion is only about half the column of wood present at the time of wounding. (Photo courtesy of Alex L. Shigo, USDA, Forest Service, Durham, New Hampshire.)

Fig. 13.4   Hollow formed in lower section (right) behind 50-year-old wound; upper section (left) contains extensive decay. In both sections, however, defect has not invaded tissues formed after woundings. (Photo courtesy of Alex L. Shigo, USDA, Forest Service, Durham, New Hampshire.)

trees to live indefinitely despite serious wounds and invading microorganisms (Fig. 13.5).

A layer of callus growth is formed at the edges of the wound in an attempt to close the wound and seal the affected area. The callus ridge is the most easily recognized expression of compartmentalization. The callus is formed. after wounding and is not invaded by microorganisms (Fig. 13.6). When the callus has closed the wound little additional discoloration and decay will occur.

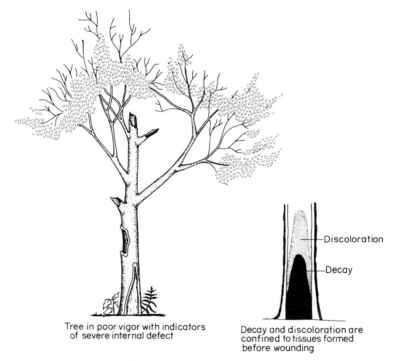

Tree in poor vigor with indicators
of severe internal defect

Decay and discoloration are
confined to tissues formed
before wounding

Fig. 13.5 Compartmentalization in a tree of low vigor.

MICROBIAL SUCCESSIONS AFTER WOUNDING

Many species of fungi and bacteria grow on wounded tissues. Some merely grow on dead exposed tissues while others invade the living tree. In most trees there is a succession of microorganisms that occurs from the initial wound to the decomposition or decay. The wound is first invaded by microorganisms, both bacteria and fungi, that attempt to break down some of the host response barriers. These microorganisms are associated with a darkening or discoloration of the wood. A second wave of microorganisms may follow, the decay fungi, which decompose the wood that was discolored by the first wave or the pioneer microorganisms. These processes may stop at any point from early discoloration next to the wound and need not proceed to decay. In most small or minor wounds the processes usually stop at discoloration when the wound is closed. This interaction of several microorganisms to cause discoloration and decay is called *microbial succession*. It is easy to see the advantage to the microorganisms to

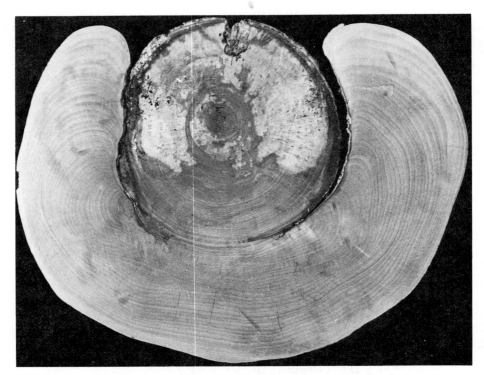

Fig. 13.6 Wound closure around a severe mechanical wound. Although decay is well advanced in the center of the tree the outside callus ridge remains free of discoloration and decay. (Photo courtesy of Alex L. Shigo, USDA, Forest Service, Durham, New Hampshire.)

employ several specialized organisms to overcome the host's defense. It is important to understand that decay usually results from an ordered sequence of events that follow wounding since attempts to control or prevent decay must be aimed at stopping or retarding this sequence in the early stages if they are to succeed.

WOUND TREATMENT

Pruning

Pruning involves the removal of dead and dying branches, as well as broken branches. It is necessary for the health of the tree but creates wounds. Branch pruning is also done to remove branches near buildings and utility lines, and for aesthetic purposes. There are three objectives of proper pruning: (1) to remove a branch with minimal injury to the tree, (2) to ensure rapid callus closure of the wound, and (3) to allow the tree to compartmentalize the injured woody tissues.

Branch pruning can be separated into removing (a) live or recently injured branches, and (b) dead branches or branch stubs. Removal in the former group can be further separated into large branches [greater than 1 inch (2.5 cm) in diameter] and small branches (less than 1 inch in diameter). Large branches must initially be undercut at least $\frac{1}{3}$ of the way through the diameter of the branch at a point approximately 1 foot (30 cm) from the trunk or from the next largest branch (Fig. 13.7). The next cut, which will sever the main body of the branch, should be made on the top of the branch a few inches (5–10 cm) farther from the trunk than the undercut. This will allow the branch to break off cleanly from the tree without tearing bark from the underside of the branch. The remaining stub should then be cut to the trunk or to the next branch. The "natural

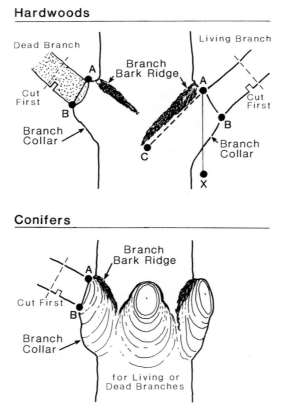

Fig. 13.7 Natural target pruning. (Photo courtesy of USDA Forest Service.)

target pruning" technique developed by Dr. Alex L. Shigo is recommended as the proper pruning method for all pruning cuts. Small branches may be severed with one pruning cut, but the weight of the branch still should be supported until it is completely severed to avoid tearing of the bark below the wound. Branches dead for several years present a unique problem in pruning because they usually have formed considerable callus around the branch stub. These branches should be severed at the point where the dead branch and the callus meet. Cutting into the callus will enlarge the wound and break the barrier compartment formed by the tree. This will delay wound closure and will also increase chances of decay formation behind the wound.

Bark Wounds

Wounds that tear the bark from the underlying wood and sometimes break wood fibers commonly occur to the trunk, roots, and branches. Bark wounds are also treated according to the age of the wound. Loose bark is removed from around both recent wounds and old wounds (Fig. 13.8). Tight bark is also removed around a recent wound with cutting tools, such as a pruning knife, until healthy bark tissue is reached. Cutting tools should be surface-sterilized after each cut with 70% alcohol or 20% household bleach. The wound is then shaped into a vertical ellipse. This will allow callus closure of the wound to occur in minimum time. Callus has often formed around most of an old bark wound but may be hidden by loose bark. Bark around older wounds is removed until a callus layer is found. Do not cut into the callus or attempt to shape the wound. This will only break the barrier zone and expand the wound. If callus is absent in any portion of the wound, remove the dead bark back to the living tissue in that portion in a manner similar to that used for recent wounds.

Wound Dressings

Although commercial wound paints are usually applied over the surfaces of both pruning and bark wounds to prevent internal decay and as an aid in wound healing, there is no experimental evidence that presently available wound paints stop decay. Research by Shigo and Wilson on wound dressings (1971, 1972) has shown no difference in internal discoloration and decay behind the wound whether dressings were applied or not. Neely (1970) found the rate of wound closure was also not affected by application of wound dressings. Periodic reapplication of dressings has sometimes been recommended in some cases to increase their effectiveness because the initial coat of dressing often cracks or peels within a few months of application. However, Marshall and Waterman (1948) found that thick coatings of asphalt wound dressing blistered and in some cases accelerated rather than retarded decay.

Why don't wound dressings that have been in such widespread use in ar-

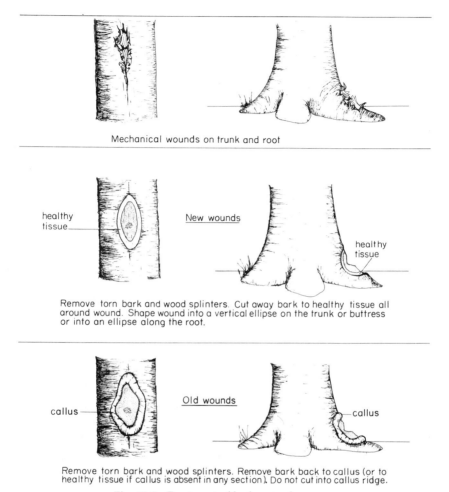

Mechanical wounds on trunk and root

healthy tissue

New wounds

healthy tissue

Remove torn bark and wood splinters. Cut away bark to healthy tissue all around wound. Shape wound into a vertical ellipse on the trunk or buttress or into an ellipse along the root.

callus

Old wounds

callus

Remove torn bark and wood splinters. Remove bark back to callus (or to healthy tissue if callus is absent in any section). Do not cut into callus ridge.

Fig. 13.8 Treatment of bark wounds.

boriculture prevent decay in trees? The answer may lie in the major emphasis that has been placed upon decay itself and decay fungi when studying wounds. As stated earlier, decay is the end result of a complex series of events that begins with a wound. Most wound dressings are materials designed to inhibit wood decay fungi, especially those that attack wood in service. Decay-producing microorganisms, however, are most important near the end of this process. Needed most are materials that will inhibit the early invading microorganisms and materials that will stimulate the host response to block the invasion of these microorganisms soon after the wound occurs. Some research using biological control against invading decay fungi has already demonstrated its effectiveness (Blanchette and Sharon, 1975; Pottle and Shigo, 1975).

There are many people, especially homeowners, who still feel that wound painting is an essential part of sound shade tree maintenance. It is important for the arborist to explain to the homeowner why wound dressings were not applied so that the arborist is not thought to be incompetent. Wound dressings should be applied if the homeowner insists on it for aesthetic purposes because little if any delay in wound closure or increase in internal discoloration and decay is caused by single thin applications of most commercially available wound dressings.

Cavity Treatment

Arborists have long sought to be able successfully to control internal decay in trees in the same way that a dentist can control decay in a tooth. The dentist working in a relatively small area under sterile conditions can remove all the decay from the tooth. She can then replace it with a filling that allows normal use and prevents recurrence of decay in the tooth. There are two problems, however, that occur when the arborist attempts on a living tree what a dentist does with a decayed tooth. First, a tree is very much larger than a tooth and creating sterile conditions inside a tree is virtually impossible. The defective tissue in the tree, both discoloration and decay, also extends far beyond any external opening and the removal of all this tissue is also not feasible. Second, the dentist must remove decay from a tooth to save it because it is the only organ in the body that cannot repair itself, while trees have systems of protection and repair that enable them to confine or compartmentalize defective tissue.

Control of decay is not the only reason that internal decay or cavities are treated. One major argument for replacing decayed wood and cavities with some type of filler material is to strengthen a tree that has obviously lost some of its structural support. Another major argument is to create a surface on the outside of a cavity so the callus will form on the outside of the tree instead of rolling into the cavity. Both these arguments, however, are currently under debate by many tree physiologists, plant pathologists, and arborists. Many of these people now feel the rigid materials like cement, in common use for cavity filling, are of little structural support because they are not flexible enough to move with the tree. The use of flexible materials like polyurethane foam, which act primarily as a surface for callus, has been suggested instead of the use of material to support the tree (King *et al.* 1970). Some tree researchers even believe that a hollow cavity with turned-in callus is stronger than a filled cavity, and that cavity treatment is detrimental to the strength of the tree.

Although there is little hope of either stopping or eliminating decay with cavity treatment, and there is some doubt as to its benefits in increasing the stability of the tree, there is still a demand for cavity treatment for aesthetic purposes on highly valuable shade trees. If cavity treatment is to be performed care must be exercised not to upset the barriers that have been formed by the tree against the further spread of decay. It is often suggested that all discolored and decayed wood around a cavity be removed until healthy wood is reached. This practice will break the barrier zone formed at the cambium when the tree was originally wounded, often many years before, and allow the decay to expand to healthy tissue (Fig. 13.9). The hard wall of a cavity with clear healthy wood, just to the outside, is evidence that the barrier zone is working and the decay is confined to tissues formed before wounding.

When treating a cavity carefully remove only the decayed wood as far into the cavity as can be reached (Fig. 13.9). Stop when sound discolored wood is reached. Cutting into discolored wood may break the barrier zone. Do not cut into the callus because this will also break the barrier zone and also delay wound closure. The cleaned cavity can then be filled and the exterior made as aesthetically pleasing as possible. Application of wound paints on the interior of the cavity before filling are sometimes recommended, but there is no evidence that they will be of any benefit to the tree.

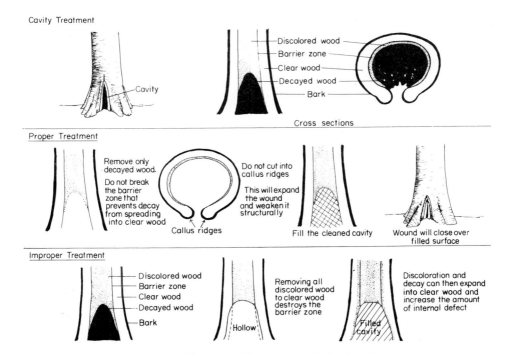

Fig. 13.9 Proper and improper cavity treatment.

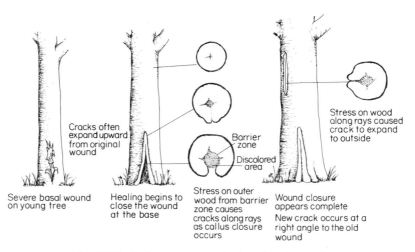

Cracks often expand upward from original wound

Barrier zone

Discolored area

Stress on wood along rays caused crack to expand to outside

Severe basal wound on young tree

Healing begins to close the wound at the base

Stress on outer wood from barrier zone causes cracks along rays as callus closure occurs

Wound closure appears complete
New crack occurs at a right angle to the old wound

Fig. 13.10   Bark seams or cracks from basal wounds.

OTHER CONDITIONS RESULTING FROM WOUNDING

Bark Seams or Cracks

Although splits along the trunks of trees are usually attributed to frost or cold temperatures, most are caused by weakness in wood from internal discoloration and decay. The affected trees usually incurred a basal wound many years ago which completely healed or is quite inconspicuous (Fig. 13.10). As the callus formed over the wound, stress resulted in the newly formed wood. Fracture zones along rays formed above the wound and eventually reached the outside of the tree. Another source of bark seams is trunk wounds that develop zones of weakness above and below the wound during callus formation (Fig. 13.11).

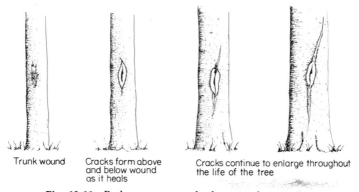

Trunk wound

Cracks form above and below wound as it heals

Cracks continue to enlarge throughout the life of the tree

Fig. 13.11   Bark seams or cracks from trunk wounds.

Fig. 13.12 Bark seams on paper birch above and below a trunk wound. (Photo courtesy of Alex L. Shigo, USDA, Forest Service, Durham, New Hampshire.)

These areas split up and down the bark and often continue to enlarge for the life of the tree (Fig. 13.12).

Tapping Wounds

Sugar maples, and in some cases other maple species, are tapped each year in the early spring for the production of maple syrup and maple sugar products from the sap. Although this process is usually conducted in a commercial sugar-bush, tapping trees for sap has also become increasingly popular for roadside and backyard sugar maples. When performed carefully and according to tree diameter guidelines for the proper numbers of taps per tree, tapping does little harm to a healthy, vigorous tree. The tapping wounds usually heal over com-

TABLE 13.1

Recommended Number of Taps for Various Diameters of Trees

Tree diameter at 4.5 feet	Tree diameter at 1.4 m	Number of taps
0 to 10 in.	0 to 25 cm	0
10 to 14 in.	25 to 35 cm	1
14 to 18 in.	35 to 45 cm	2
18 to 22 in.	45 to 55 cm	3
Greater than 22 in.	Greater than 55 cm	4

pletely during the growing season. However, many inexperienced homeowners may cause severe injury to valuable shade trees during the drilling of tapholes, the placement and removal of taps, and especially from greatly exceeding the recommended number of taps per tree. The generally accepted guidelines are as given in Table 13.1.

In addition, many sugar maples growing along roads and around homes are not healthy, vigorous trees. In fact, many are suffering from progressive degrees of maple decline (see Chapter 23, Diebacks and Declines). These trees are much more severely injured by tapping than trees in a sugarbush and are much more susceptible to internal discoloration and decay, as well as to an accelerated decline in vigor. Only shade trees in excellent condition should be selected for tapping, and it should be conducted according to the guidelines developed by the maple sugar industry.

Tree Injection Wounds

Injection of materials into the trunks of trees, either under pressure or by gravity, is a widely used practice in arboriculture. Nutrients, fungicides, insecticides, antibiotics, and growth regulators are some of the chemicals that are injected into trees. Although there are many benefits to trees from proper injections, wounds made for tree injections have always been a concern for those who manage the long-term care of trees. The injection wounds used to introduce these materials into the trunk have varied in wound diameter and depth. In general, the larger the diameter of the injection hole and the deeper the hole is made, the more serious that wound is to the tree.

Unfortunately, large and deep wounds have often been employed to inject trees, especially American elms, since large volumes of material could be forced into the trunk through such large wounds. Injection deep into the tree, however, does not result in materials being evenly distributed within the crown of the tree as desired. High pressure injection also does not help to distribute the injected materials to the crown where they are needed, but sometimes can seriously damage a tree.

Most of the active transport within a tree occurs in the current year's wood. If materials to be injected are introduced through shallow wounds that are drilled just into the current year's wood, maximum uptake and distribution can be achieved. Using this type of injection, only very low pressure (approx. 1 atmosphere) need be employed. In addition, there is no need to use large-diameter holes to introduce materials into the tree. The use of shallow injection holes $\frac{3}{16}$ inch (5 mm) in diameter, placed every 6 inches (15 cm) around the trunk at ground line, will achieve more effective uptake and distribution than a similar system employing large and deep wounds. Shallow and small-diameter injection holes, sometimes referred to as "microinjection," usually close completely within the season of injection. Large and deep injection holes, sometimes referred to as "macroinjection," greatly increase the chances of slow wound closure and the development of internal discoloration and decay.

The production of ooze from the injection hole, or bark dieback or vertical cracks around it are danger signs. If any of these symptoms occurs at several injection wounds, no further injections should be performed on that tree for at least one season.

Each injection wound must be made carefully. Always use sharp drill bits. Insertion and removal of injectors should be done with minimal bark injury.

Tree implants are placed into trees through wounds similar to those used in tree injection. Concerns similar to those about the size and depth of injection wounds apply to implant wounds. Implants, in contrast to injections, use dry materials that are absorbed into the sap of the tree. Implants are made of plastic and are not designed to be removed once inserted into the wound. The top of the implant must be placed even with the bark–wood interface (cambium area) to allow callus closure. If the implant is improperly placed above this area the callus will often pinch around the top of the implant and delay wound closure. If oozing, bark dieback, or bark cracking occur to several implant wounds, avoid further implants on that tree for at least one season.

Injection holes are sometimes plugged with wooden dowels after treatment in hopes that decay-producing microorganisms will be prevented from entering the plugged wounds. However, no differences were found in the amount of decay that formed following wounding in studies where similar wounds were either plugged or left open.

Slime Flux

Liquid ooze or flux sometimes is produced from wounds. These exudations, called slime flux, are frequently foul-smelling and discolor the bark below the wound (Fig. 13.13). Fluxing from wounds is common in a large number of species including apple, beech, birch, hickory, linden, maple, oak, poplar, sycamore, and willow, but is most common in elms. The flux may be initially clear as it leaves the wound but becomes slimy as it dries and is colonized by numerous saprophytic microorganisms. It may color the bark a variety of colors, the most common being white, brown, gray, or black.

Fig. 13.13   Slime flux on bark (A) below a branch wound on American elm and (B) below a weak fork on American beech. (Photo A courtesy of Shade Tree Laboratories, University of Massachusetts, Amherst.)

Slime flux is caused by the production of gas by the metabolic activities of bacteria (fermentation) in columns of discoloration and decay that follow wounds (Fig. 13.14). Fluids within the column are forced out of the wound by internal gas pressure. The nature of the slime flux is determined by the microbial flora within the column.

In most cases slime flux appears to have few detrimental effects on the tree's health; however, sometimes production of slime flux inhibits callus closure of wounds. Control of slime flux involves drilling into the column of defect to relieve the pressure and inserting a perforated pipe into the hole to allow continual drainage away from the tree. Procedures for such treatment of slime flux are outlined in detail by Carter (1969). However, these procedures do not cure slime flux and create yet another wound that breaks the compartment and thereby allows discoloration and decay to spread outside the column.

SPROUT CLUMP TREATMENT

Dense groupings of trees are commonly found coming from one common base. These trees are referred to as *sprout clumps* because they originated as sprouts around the base of a parent tree after it was cut. Sprout clumps are common around homes in wooded lots because logging is often done before the

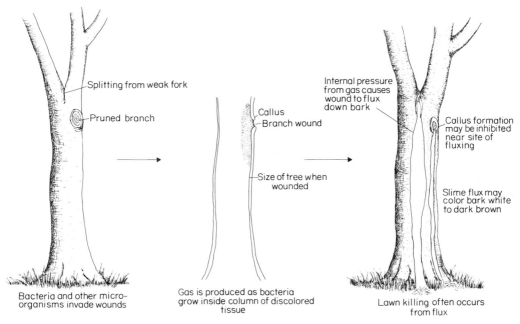

Fig. 13.14 Slime flux.

land is sold for development into house lots. The sprouts originate from dormant buds in the bark around the root collar that become active when the tree is cut. Many of these sprouts, however, develop from a single tree and there is a large amount of competition between them. Often the entire clump is in poor vigor with many poorly healed branch stubs and several dead and dying sprouts. In most cases the sprout clump must be treated by removing most of its members if any valuable shade trees are to result. Selection of the one or two

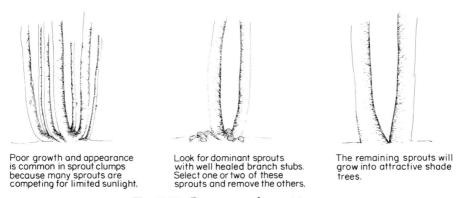

Fig. 13.15 Proper care of sprout trees.

stems that are to be kept should be based upon vigor and wound healing (Fig. 13.15). Large stems with well-healed branch stubs 4 to 12 feet (1 to 4 m) above-ground are the best choices. The remaining sprout(s) will usually develop into shade trees that become a valuable asset to the homeowner.

LITERATURE CITED

Blanchette, R. A., and E. M. Sharon. (1975). *Agrobacterium tumefaciens*, a promotor of wound healing in Betula alleghaniensis. *Can. J. For. Res.* **4**, 722–730.

Carter, J. C. (1969). The wetwood disease of elm. *Ill. Natl. Hist. Surv., Circ.* **50**, 1–19.

King, G., C. Beatty, and M. McKenzie. (1970). Polyurethane for filling tree cavities. *Univ. Mass., Co-op. Ext. Serv., Publ.* No. 58, pp. 1–6.

Marshall, R., and A. Waterman. (1948). Common diseases of important shade trees. *U.S., Dep. Agric., Farmers' Bull.* **1896**, 1–34.

Neely, D. (1970). Healing of wounds on trees. *Am. J. Hortic. Sci.* **95**, 536–540.

Pottle, H. W., and A. L. Shigo. (1975). Treatment of wounds on Acer rubrun with Trichoderma viride. *Eur. J. For. Res.* **5**, 274–279.

Shigo, A. L., and C. L. Wilson. (1971). Are wound dressings beneficial? *Arborist's News* **36**, 85–88.

Shigo, A. L., and C. L. Wilson. (1972). Discoloration associated with wounds one year after application of wound dressings. *Arborist's News* **37**, 121–124.

SUGGESTED REFERENCES

Houston, D. R. (1971). Discoloration and decay in red maple and yellow birch: reduction through wound treatment. *For. Sci.* **17**, 402–406.

Shigo, A. L. (1973). A tree hurts, too. *U.S., For. Serv., Res. Pap.* **NE-INF-16-73**, 1–28.

Shigo, A. L. (1975). Some new ideas in tree care. *J. Arbori.* **1**, 234–237.

Shigo, A. L. (1976). Rx for wounded trees. *U.S., For. Serv.* **A1B-387**, 1–37.

Shigo, A. L. (1987). A new tree biology. Shigo & Associates, Durham, NH. 595p.

Shigo, A. L., and E. H. Larson. (1969). A photo guide to the patterns of discoloration and decay in living hardwood trees. *U.S., For. Serv., Res. Pap.* **NE-127**, 1–100.

Shigo, A. L., and F. M. Laing. (1970). Some effects of paraformaldehyde on wood surrounding tapholes in sugar maple trees. *U.S., For. Serv., Res. Pap.* NE **NE-161**, 1–11.

Shigo, A. L., and W. C. Shortle. (1977). "New" ideas in tree care. *J. Arbori.* **3**, 1–6.

Tattar, T. A. (1973). Management of sprout red maple to minimize defects. North. Logger. 21: 20–21.

Wilson, C. L., and A. L. Shigo. (1973). Dispelling myths in arboriculture today. *Am. Nurseryman* **127**, 24–28.

PART II

NONINFECTIOUS DISEASES

14

Introduction to Noninfectious Diseases

In addition to the common living pathogens—fungi, bacteria, viruses, mycoplasmas, nematodes, and parasitic seed plants—there is a large number of nonliving (abiotic) agents that cause disease in plants. Most of these nonliving agents are stress factors that adversely affect the health of the plant. These health problems are usually termed *noninfectious diseases* because in contrast to disease associated with living pathogens they cannot spread from plant to plant.

Noninfectious diseases are often the most common group of tree diseases with which the shade tree specialist comes in contact, and they are among the most troublesome problems. Why are noninfectious diseases so common in shade trees? A better understanding of the basic needs of trees is needed to completely answer the question.

SOME COMMON BASIC NEEDS OF TREES

1. A balance must be maintained between water lost through the leaves and that taken up through the roots.
2. Soil conditions must allow sufficient growth of roots to be able to supply the crown with moisture and nutrients.
3. Photosynthesis must be able to continue at a rate sufficient to supply the energy needs of the tree, and its products must be conveyed from the leaves to the stem and roots.
4. Vigor of the tree must remain high enough to prevent attack by weak infectious pathogens and by secondary insects.

If any one of the basic needs is missing, that factor or its inhibiting agent can become a noninfectious agent.

SOME REASONS FOR PREVALENCE
OF NONINFECTIOUS DISEASE

1. Shade trees are long-lived individuals in a *rapidly changing* environment (Fig. 14.1). The amount of stress placed upon a tree is directly related to the rate of change of its microenvironment.

2. Shade trees are often planted in suboptimum locations. Roadsides and lawns are often exposed areas with poor soil conditions and too frequent interaction with people (Fig. 14.2). Competition with lawns for moisture and nutrients usually ends with the tree getting less than it needs for adequate growth.

3. Many varieties of ornamental trees are selected by nurseries, whenever possible, for resistance to infectious diseases. Planting of species or varieties known to have troublesome disease problems is discouraged. Therefore, many infectious diseases can be avoided.

4. Noninfectious diseases attack all species!

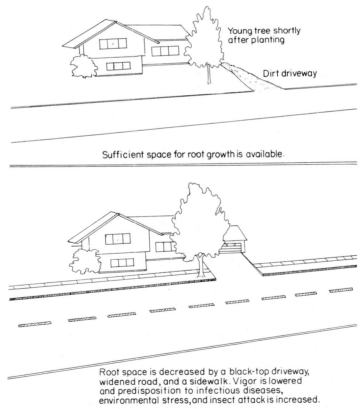

Young tree shortly after planting

Dirt driveway

Sufficient space for root growth is available.

Root space is decreased by a black-top driveway, widened road, and a sidewalk. Vigor is lowered and predisposition to infectious diseases, environmental stress, and insect attack is increased.

Fig. 14.1 Trees in a rapidly changing environment.

Fig. 14.2 Roadside sugar maple in decline from numerous interactions with the activities of people including injury during construction and from lawnmowers.

GENERAL SYMPTOMATOLOGY OF INFECTIOUS AND NONINFECTIOUS DISEASES

Noninfectious diseases as a general rule have uniform symptoms on the entire tree while the symptoms of infectious diseases on the tree are variable and often random in severity. Noninfectious diseases often affect every tree of the same species in a small area with identical symptoms, and some noninfectious diseases will affect more than one species. These last two characteristics can rarely be associated with infectious diseases. Most infectious diseases, on the other hand, will attack trees of one species at random or in a progressive sequence in a particular location. Infectious diseases rarely attack all the trees in an area at the same time, and infectious diseases rarely affect more than one species in the same manner. In fact, most infectious diseases are species-specific, often facilitating diagnosis once the host species and typical symptoms are correctly identified, while most noninfectious diseases are nonspecies-specific and often affect all woody and nonwoody plants to some degree in a particular area. Regularity of symptom expression and the lack of any evidence of living pathogens are the best diagnostic criteria for the noninfectious diseases.

CATEGORIES OF NONINFECTIOUS DISEASES

There are many varied forms of detrimental stress that affect shade trees but most noninfectious diseases fall into the following three broad categories: environmental stress, animal injury, and people pressure. There is, of course, a great deal of overlap between and within these categories, and there also are many cases where several noninfectious diseases may interact simultaneously on a tree and also interact with one or more infectious diseases. Environmental stress results most commonly from extremes of temperature and moisture, and from a soil environment that does not allow normal root function. Animal injury can result from the activities of dogs, cats, birds, horses, rabbits, squirrels, and many other pets and wildlife around shade trees. People pressure results from the diverse activities of man that are detrimental to trees. Construction of buildings and roads; use of deicing salt; misuse of agricultural chemicals; improper pruning, planting, and wound treatment; soil compaction; and air pollution are some common forms of people-pressure diseases on shade trees.

SUGGESTED REFERENCES

Roberts, B. R. (1977). The response of urban trees to abiotic stress. *J. Arbori.* **3,** 75–78.
Wilson, C. L. (1977). Emerging tree diseases in urban ecosystems. *J. Arbori.* **3,** 69–71.

15

Temperature Stress

INTRODUCTION

Extremes of temperatures, as well as rapid changes in temperature, can be injurious to trees. Shade trees are often exposed to larger variations in temperature than trees growing in the forest and are much more susceptible to temperature injury. In many cases there is also a considerable interaction between temperature and moisture, which will be discussed in the following chapter on moisture stress. The effects of temperature are most severe on newly planted trees that have not adjusted yet to their new environment. Trees whose environments have changed around them, usually due to the activities of man, are also likely to be injured by extremes and rapid changes in temperature. Since many shade trees often fall into these two categories, temperature injury is often a concern of the arborist.

HIGH TEMPERATURE

Summer and Winter Sunscald

Whenever trees growing in deeply shaded locations are suddenly exposed to intense sunlight, stress results on the newly exposed bark from rapid temperature increases. This happens whenever a forested area is excessively thinned to create a house lot or recreation area and whenever trees are moved from a shaded nursery to an exposed location, such as the roadside or lawn. Two events may occur as a result of this type of stress, summer sunscald and winter sunscald. Summer sunscald is heat injury to the exposed bark during the summer and results in bark killing with subsequent canker formation (Fig. 15.1). Wood beneath the dead bark is sometimes invaded by decay fungi and trees may break at this area a few years later (Fig. 15.2). Summer sunscald injury, combined often with accompanying drying of the sites, has accounted for a large number of tree

Fig. 15.1 Summer sunscald injury on black birches near the edge of an area cleared for house construction. (A) Recent injury; (B) old injury. Note exposed wood on injured face and callus roll at edges.

losses around new homes and in newly created recreation areas. Winter sunscald is injury from rapid changes in bark temperature during cold and sunny winter days. Exposed bark, especially on species with dark bark, becomes much warmer on the sunny side than the air during these days but cools very rapidly after sunset. The rapid temperature changes can result in bark injury that usually occurs in the southwest side of the tree. Bark temperature reaches its maximum in midafternoon when the sun is in the southwest quadrant and, consequently, injury usually occurs most severely in that section of the tree (Fig. 15.3). The other symptoms of winter sunscald injury are similar to those found in summer sunscald except that most trees, although often severely cankered on the southwest side, usually survive the injury.

Fig. 15.2 Tree breakage from decay following summer sunscald injury on black birch.

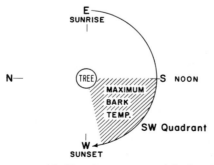

Fig. 15.3 Winter sunscald. (Drawing courtesy of the late Donald Curtis.)

Summer sunscald and winter sunscald can both be prevented in newly transplanted or recently exposed trees by wrapping trunks of trees with protective materials like Kraft paper from the ground to the first branches. Summer sunscald may also require watering to relieve accompanying moisture stress (see Chapter 16, Moisture Stress). The light colored wrap prevents excessive heating of the bark and should be left in place for at least 2 years. Most trees can adapt to increased exposure to the sun by producing a thicker bark, but it takes them about 2 years to accomplish these changes. This is why trees that have always grown in exposed locations do not need continuous protection from wraps. Wrapping trees, of course, is part of the proper transplanting procedure for trees and should be used even in cases where the danger of temperature injury may appear slight. Recently, the application of white water-base paint has also been found effective when applied to exposed portions of the trunk and branches. However, a study by Mosher and Cool (1974) found some latex paints caused Cytospora cankers. This method may be undesirable to some homeowners because the paint remains on the trees for several years.

Heat Injury

Injury due to excessively high summer temperatures is a problem mainly on young nursery trees and is found most commonly in the southern and southwestern United States. However, if any young stock from the greenhouse is suddenly exposed to intense heat, injury can occur. Partial shade from overhead slats can decrease the chances of heat injury until the plants adjust to the summer conditions.

LOW TEMPERATURE

One of the major barriers to the geographic movement of plant materials by man is cold temperature or frost injury. Native trees that have adapted to northern temperature climates are not usually injured by low temperatures. Exotic trees from more southern latitudes have not adapted to the temperature peculiarities of particular locations and are usually the most prone to cold temperature injury.

Woody plants have adapted to winter conditions by an established pattern of growth and dormancy that follows the yearly weather cycle very closely. They can tolerate extreme cold during the winter but little during the growing season. This phenomenon is termed "hardiness." As fall approaches woody plants begin to become progressively more cold hardy, and reach a peak of hardiness in midwinter. Plants begin to decrease in hardiness during early spring and reach a low point of cold tolerance during the spring flush. This is the most vulnerable time for cold injury. A late spring frost can do considerable damage to many trees and can even kill whole trees (Fig. 15.4). Injury is most commonly seen on

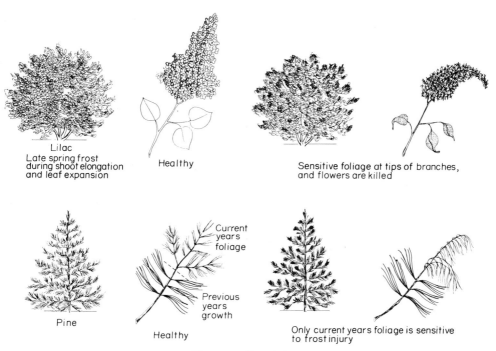

Lilac
Late spring frost
during shoot elongation
and leaf expansion

Healthy

Sensitive foliage at tips of branches,
and flowers are killed

Pine

Healthy

Current
years
foliage

Previous
years
growth

Only current years foliage is sensitive
to frost injury

Fig. 15.4   Frost injury.

flowering trees, such as crabapples, magnolias, and lilacs, whose flowers are often killed by late frosts (Fig. 15.5). The later into the spring season the frost occurs the greater the chances that even native shrubs and trees will be injured.

Frost injury cannot be controlled but in most cases it could be prevented, especially in cases where species have been planted north of their cold tolerance limits. Each species of tree and shrub is assigned a hardiness zone number that corresponds with the northernmost geographical area where the species can be planted safely. It is, therefore, very important to be aware of the cold tolerance of any tree or shrub before purchasing it or planting it for a client.

Although hardiness normally increases in the fall the plant can be "fooled" by late summer (after early July) fertilizer applications. High-nitrogen fertilizers cause excessive foliage production and delay hardening. If these fertilizers are applied in late summer they can cause flushing of foliage that may not have time to harden before fall frosts begin. Balanced fertilizers ·or those not excessively high in nitrogen should be applied to trees and shrubs and summer fertilization should be avoided.

Frost injury can be recognized by a browning of newly emerged shoots and flowers on the entire tree or shrub (Fig. 15.6). However, in a frost pocket only the lower parts of the tree may be killed back (Fig. 15.7). Injury to every tree of

Fig. 15.5 (A) Magnolia flowers killed by late frost. (B) Close-up.

Fig. 15.6 Frost injury on white spruce. Note drooping of injured terminal growth.
(Photo courtesy of Darroll D. Skilling, USDA Forest Service, St. Paul, Minnesota.)

Fig. 15.7 (A) Frost injury on lower branches on a pignut hickory. (B) Recovery later in the season. (Photo courtesy of Shade Tree Laboratories, University of Massachusetts, Amherst.)

that species in a limited local area is also characteristic of frost injury. Injury occurs on evergreens as well as on hardwoods. Check with local weather bureaus for low-temperature data for the past 2 weeks if frost injury is suspected.

CONTAINERIZED ORNAMENTAL PLANTS

Roots of trees and shrubs are much more susceptible to low-temperature injury than stems. Roots are normally protected from extremely low temperatures by the insulating effect of the soil. Restricted planting space in many urban areas has prompted the widespread use of shade trees and ornamental shrubs in movable containers. Roots of trees and shrubs growing in containers aboveground, however, are often exposed to much lower temperatures. These woody plants are not insulated from the low temperatures, and root injury often occurs. Root injury, particularly of the young roots, due to cold temperatures has resulted in considerable mortality in container-grown woody plants. Both large potted shade trees as well as small shrubs are susceptible to low temperature injury when left in exposed areas during the winter. Containerized trees and shrubs can be protected from injury by moving them into an unheated greenhouse, placing them in the soil, or mulching heavily around the entire container. Potted shrubs and small trees can be moved more easily and placed into greenhouses or into the soil than can large potted trees. Occasionally, the containers holding all large trees in an area are moved together and mulched. In this manner the root balls of each tree can act as insulation for the others and less mulching is needed for each tree.

RAPID CHANGES IN TEMPERATURE

Shade trees can also be injured by sudden and large increases or decreases in temperature, especially when the changes far exceed the normal range for that season of the year. A sudden increase in temperature or "thaw" during the winter may cause premature flushing of foliage or a decrease in hardiness. The buds or emerging foliage are often killed when the temperature again returns to its normal range. Sudden decreases in temperature can cause injury even when temperatures do not reach below freezing. Many woody plants, especially tropical and subtropical trees and shrubs, are sensitive to sudden drops in temperature and may become injured when such drops occur. These plants should be grown in sheltered locations, such as close to buildings or in courtyards, to avoid large temperature changes.

LITERATURE CITED

Mosher, D. G., and R. A. Cool. (1974). Protective paint induces canker formation. *Arborist's News* **39**, 42–44.

SUGGESTED REFERENCES

Evert, P. L. (1967). The physiology and cold hardiness in trees. *Proc. Int. Shade Tree Conf.* **43**, 40–50.
Good, G. L., P. L. Steponkus, and S. C. Wiest. (1976). Winter protection of containerized ornamental plants. *J. Arbori.* **2**, 51–52.
Levitt, J. (1972). Responses of plants to environmental stresses. Academic Press, New York.

16

Moisture Stress

INTRODUCTION

Moisture is a critical component of plant health. Most key plant functions such as photosynthesis and transpiration cannot occur unless the necessary water balance is available throughout the plant. Water lost through transpiration furnishes the force that lifts water from roots to leaves. However, if there is insufficient water absorbed from the soil to replace that lost during transpiration, a water deficiency occurs. If the balance is not restored, wilting of foliage will occur within a few days. Too much moisture can also be detrimental and can cause the roots to suffocate and die. Each tree species has different capabilities to tolerate various types of soil moisture conditions.

Under natural forest conditions trees are able to adapt to periodic changes in soil moisture. Through natural selection wet sites are populated by moisture-loving tree species and dry sites are populated by species preferring well-drained conditions. Shade trees, however, are usually planted trees that may not be best suited for soil moisture conditions where they are planted. Soil around roadside and backyard trees rarely receives and holds as much moisture as soil in the forest. Shade trees often must compete with lawns for soil moisture. In many cases much of their root systems are covered with roads, sidewalks, parking lots, or some other impervious layer, which causes most rain water to run off instead of entering the soil. In most urban areas the soil is laced with drainage pipes that carry moisture away from tree roots before much of it can be absorbed. It is, therefore, not surprising that moisture stress is a common problem in shade trees. Most moisture stress falls into one of several categories: moisture deficiency, moisture excess, erosion, and physical damage from snow and ice, each of which will be discussed in detail.

WATER DEFICIENCY

Drought

Trees are subject to two kinds of water deficiency stress: (1) short-term drought during one growing season, and (2) long-term drought that accumulates moisture stress over more than one growing season. The latter is the most important to trees because, in contrast to annual crop plants, trees are sensitive to year-round moisture conditions.

An important contributing factor to moisture stress is, of course, subnormal rain and snowfall. There are, however, many site factors that increase the susceptibility of shade and ornamental trees to moisture stress. Restricted root space is probably one of the most important contributing factors to moisture deficiency stress. In many cases, trees growing in confined locations, such as trees along a street, are sandwiched between roads, sidewalks, and driveways. These trees often are not able to extend their roots into sufficient soil area for them to meet the demands for moisture of the crown. They can usually survive under normal moisture conditions by growing at a slow rate, but are usually the first to be affected by drought. Trees in wells, along roadsides, on ledges, or in shallow soil are also in locations prone to moisture stress and can also be considered drought prone. Trees whose roots are shallow because of high water tables (such as trees next to rivers and lakes) are also quite susceptible to drought when the water table falls. Often it is the trees nearest bodies of water that, ironically are the first to die in a severe drought.

One of the most common symptoms of moisture stress is leaf scorch (Fig. 16.1). This symptom involves the browning of the leaf margins of hardwoods (Fig. 16.2) and of the needle tips of conifers. Scorch can sometimes be confused on hardwoods with anthracnose in its earlier stages, but leaf scorch rarely crosses over the leaf veins on the inside of the leaf while vein crossing is common in anthracnose. Leaf scorch is often most severe in the upper branches while anthracnose is usually most severe nearest to the ground. Leaf scorch is caused by large losses of moisture by the leaves from evaporation that cannot be restored by the roots due to low soil moisture. The edges of the leaves that are farthest from main veins dehydrate and die, giving the leaf a burnt or scorched appearance.

Another common symptom of moisture deficiency is interveinal necrosis. The leaf tissues around the major veins remain green but the tissue between the veins turns brown. Since the veins carry the moisture to the leaf the tissue adjacent to them is the last to suffer from water deficiency. In addition, midsummer defoliation sometimes occurs from drought (Fig. 16.3).

If moisture deficiency exists for more than one season, crown decline will usually begin. Dieback of twigs and small branches in the upper crown usually occurs first, and later progressively larger branches are killed. Sucker growth may also appear on the main trunk and large branches of some species. There is

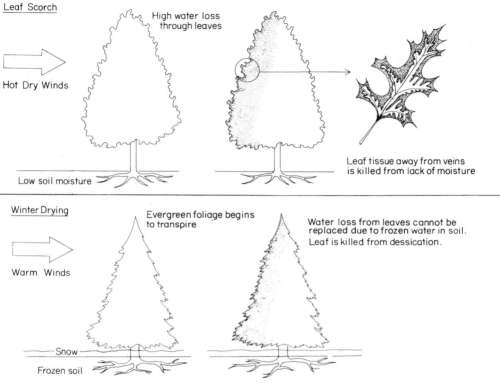

Leaf Scorch

High water loss through leaves

Hot Dry Winds

Low soil moisture

Leaf tissue away from veins is killed from lack of moisture

Winter Drying

Evergreen foliage begins to transpire

Warm Winds

Water loss from leaves cannot be replaced due to frozen water in soil. Leaf is killed from dessication.

Snow

Frozen soil

Fig. 16.1 Leaf scorch and winter drying.

often a delay between the drought period and the time of symptom expression that causes problems in diagnosis of moisture stress. Since water deficiency may cause extensive root injury in late summer and fall, the current year's foliage may not reveal any symptoms. However, the effects of the previous year's drought may occur on trees the following year, when rainfall may be normal. It is, therefore, important to be aware of yearly trends in rainfall.

The symptoms of moisture stress usually do not appear until late in the summer after extensive periods of hot and windy dry weather. Treatment for moisture stress, however, should begin in advance of symptom expression. In many species, particularly the conifers, drought symptoms indicate a tree already in dangerously poor health.

Although no trees are immune to drought, those growing in the forest or in forestlike conditions are more resistant to moisture stress than trees growing along streets and with lawn over their roots. Trees with unrestricted root space and trees with their roots covered with natural or added mulches are least likely to suffer from periods of low precipitation. Not all trees, of course, can have these ideal conditions because the environment of our roadsides, where trees are

Fig. 16.2 Leaf scorch from drought on (A) chestnut oak; (B) flowering dogwood. (Photo courtesy of Shade Tree Laboratories, University of Massachusetts, Amherst.)

Fig. 16.3 Midsummer defoliation of sugar maple and American elm trees from drought. (Photo courtesy of Shade Tree Laboratories, University of Massachusetts, Amherst.)

needed most, cannot be changed back completely into a forest setting. It is possible, however, to modify the roadside environment by adding thick mulches around newly planted and existing trees to minimize moisture stress and by planting clumps of trees or wide tree belts that can create forestlike conditions.

If road widening and sidewalk straightening and leveling could be avoided, large numbers of existing shade trees would be saved from loss of even more root area. Selecting tree species that can adapt to limited moisture and avoiding rapid-growing trees that soon outgrow their available root area are practices that would decrease the moisture-stress problem on shade trees along roadsides. Similar problems exist in privately owned yard trees but, in this case, watering of a few trees can be prescribed to correct the problem. Watering should be done over roughly a 12-hour period with a slow running hose that is periodically moved to thoroughly wet down the entire root space to a depth of 12 to 18 inches (30 to 50 cm). Watering should be done not sooner than 10 days after the last substantial rainfall. Newly transplanted trees, of course, need special attention and will be discussed in depth in Chapter 21, Tree Maintenance.

Wetting Agents

Water, whether from rain or from irrigation, often does not penetrate the soil evenly, resulting in wet and dry areas of soil around the roots of trees and shrubs. Wetting agents have been used for many years in the turf industry to break the surface tension of soils and achieve even soil wetting. Recent studies with newly planted trees and shrubs in New York City's Central Park have shown that plants treated with a wetting agent were better able to tolerate drought stress than untreated plants (Tattar, 1986). It is suspected that the wetting agent may have allowed greater root growth. In Tattar's study, terminal twig growth, which usually balances root growth, was measured and found to be substantially greater in the plants treated with the wetting agent than in untreated plants. Since moisture management is critical to the survival and growth of woody plants, landscape managers should consider the use of wetting agents in moisture problem sites.

Winter Drying

Broadleaved and needled evergreens are subject to loss of water during the winter. Since the soil around the roots is normally frozen, water lost through transpiration at this time cannot be replaced. Severe water loss usually occurs in late winter on warm and windy days. The type and cause of injury are similar to those found in leaf scorch (Fig. 16.4) but, since this type of moisture deficiency stress usually occurs in the winter and early spring, it is called winter drying or winter burn (Fig. 16.1). Symptom expression is not evident until spring. The affected foliage, appearing yellow to dark brown, presents a sharp contrast to the new emerging foliage.

Fig. 16.4   Leaf scorch on yew caused by winter drying. (Photo courtesy of Shade
Tree Laboratories, University of Massachusetts, Amherst.)

Although all evergreens are susceptible to winter drying, some species, such
as arborvitae and rhododendron, seem to be among the most severely affected.
Winter drying occurs most commonly when evergreens, especially sensitive
species, are planted in exposed locations. Whenever a living hedge or an exposed planting of evergreens is desired, it is advised that sensitive species be
avoided. Some success in preventing winter drying has been achieved with the
use of burlap screens, which decrease the drying effects of winds, around
shrubs and trees in winter. Antitranspirants sprayed on foliage in the fall can
also be effective on exposed trees that often have symptoms of winter drying.

WATER EXCESS

Flooding and Saturated Soil

The roots of trees are obligately aerobic and need gaseous exchange of oxygen
and carbon dioxide between the air and the soil for them to survive. The concentration of oxygen in the air aboveground is usually about 21%. The oxygen
must diffuse into the soil to reach the roots. Oxygen is usually in slightly lower

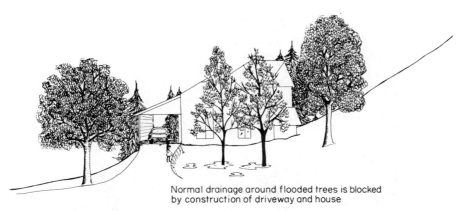

Normal drainage around flooded trees is blocked
by construction of driveway and house

Fig. 16.5 Flooding injury.

concentration (10 to 21%) in the well-aerated soil than in air, but oxygen concentration drops abruptly to about 1% near the water table. Tree roots are injured whenever the oxygen concentration drops below 10%, and root growth stops entirely at concentrations below 3%. When water stands over the roots or the soil becomes saturated for long periods during the growing season, gaseous exchange cannot take place between roots and air, and soil conditions become anaerobic. The roots suffocate under these conditions, and most trees will soon begin to decline and die (Figs. 16.5, 16.6). The effects on a tree of any given period of inundation or soil saturation will vary with the species, time of year, and duration of suffocation stress. In general, the effects of water excess (1) will be greatest during the growing season, (2) will be directly related to the duration of the stress, and (3) will occur most quickly on upland species not tolerant to natural flooding.

The best way to avoid problems with excess moisture is to determine the soil moisture and drainage patterns of an area. Avoid planting in areas that pool water after rains or that are flooded in the spring unless "water-loving" trees such as willows and poplars are planted (Table 16.1). It also may be possible to change or improve drainage patterns to eliminate flooding or saturated soil. This is especially recommended in areas where the natural drainage patterns have been altered by construction or by other activities of people.

Overwatering

Trees and shrubs should not be watered more than once a week. Since proper watering of woody plants involves the saturation of the soil to a depth of at least 12 inches (30 cm), several days are required for the roots to absorb the water and for excess water to drain away from the roots. If the soil is kept saturated by too frequent watering the roots will suffocate from lack of oxygen. Excessively wet

Fig. 16.6 Established Norway maples killed by flooding from construction-caused changes in drainage. Note standing water and wetland vegetation in foreground.

soil conditions also favor the attack of root disease fungi (see Chapter 10, Root Diseases).

EROSION

Loss of soil around roots of trees due to erosion is a common problem on steep grades and near bodies of water. Undercutting of banks with subsequent blow-down of adjacent trees is a common result of erosion. Whenever a hillside must be disturbed for a house or road, the new bank must be restored to a stable

TABLE 16.1
Tolerance of Various Tree Species to Wet
Sites and Occasional Flooding

Tolerant	Intolerant
Ash	Chestnut oak
Black gum	Eastern white pine
Cottonwood	Hemlock
Elm	Paper birch
Overcup oak	Red cedar
Pin oak	Red oak
Poplars	Red pine
Red maple	Sugar maple
River birch	White spruce
Silver maple	
Sweet gum	
Sycamore	
White cedar	
Willows	

angle. Ground vegetation should be established as soon as possible to retain soil around the remaining trees. Trees growing near water present a special problem since they have developed shallow root systems from growing near the high water table. These trees should be protected from shoreline erosion by placing a breakwater in front of exposed trees.

Fig. 16.7   Branch and trunk breakage from snow and ice. (Photo courtesy of Shade Tree Laboratories, University of Massachusetts, Amherst.)

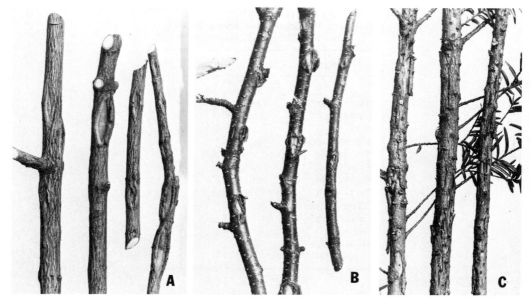

Fig. 16.8 Hail injury to (A) Norway maple; (B) white birch; (C) hemlock. (Photo courtesy of Shade Tree Laboratories, University of Massachusetts, Amherst.)

PHYSICAL DAMAGE FROM SNOW AND ICE

The extra weight of snow and ice on trees can sometimes result in considerable damage to the trees, utility lines, and property (Fig. 16.7). Although winter storms cannot be prevented, damage could be avoided by preventing weak forks, which are the most common cause of winter branch and trunk failure. Weak forks arise from branches growing at such an acute angle that normal wood formation is inhibited and a structural weakness occurs (see Chapter 26, Living Hazard Trees). Weak forks can either be eliminated early when the tree is small or secured by placing a cable between the forks. Tree species, such as silver maple, that are prone to weak forks should be avoided in locations near roads, buildings, and utility lines.

Small trees and shrubs growing close to buildings can be severely injured from falling ice and snow. Steps should be taken in cases where the danger is apparent to protect valuable foundation plantings by placing some type of strong lattice over them during the winter months.

Trees can also be injured by hail (Fig. 16.8). The force of large hailstones causes injury to the top side of branches. Severe injury, however, may girdle branches.

LITERATURE CITED

Tattar, T. A. (1986). Wetting agents benefit trees. *Grounds Maintenance* **21**, 126.

SUGGESTED REFERENCES

Butler, J. D., and B. T. Swanson. (1974). How snow, ice injury affects different trees. *Grounds Maintenance* **9**, 29–30 and 40.

Harris, R. W. (1962). Water-hazardous necessity. *Proc. Int. Shade Tree Conf.* **38**, 182–188.

Kozlowski, T. T., and W. J. Davies. (1975a). Control of water balance in transplanted trees. *J. Arbori.* **1**, 1–10.

Kozlowski, T. T., and W. J. Davies. (1975b). Control of water loss in shade trees. *J. Arbori.* **1**, 81–90.

Messenger, A. S. (1976). Root competition grass effects on trees. *J. Arbori.* **2**, 228–230.

Rusden, P. I. (1967). Drought—its effects on trees and what can be done about it. *Proc. Int. Shade Tree Conf.* **43**, 93–105.

Whitcomb, C. E., and F. C. Roberts. (1973). Competition between established tree roots and newly seeded Kentucky bluegrass. *Agron. J.* **65**, 126–129.

17

Soil Stress

INTRODUCTION

There is perhaps no environmental factor more important to the health of trees than the soil conditions in which they grow. Soil was once thought to be an inert entity, a medium containing only water and nutrients available for plant growth. Soil is now known to be an extremely complex mixture of organic and inorganic materials including a large variety of living organisms. Soil conditions, like moisture conditions, have a selective effect on plant growth. Under natural forest conditions only the best-adapted trees will grow on each soil type. In transplanted shade trees, however, the soil type may be unsatisfactory for a particular tree. Soil stress often results from this type of incompatibility, and the planted tree may not survive or may grow so poorly as never to become an attractive shade tree. Knowledge of the soil conditions that favor each tree and shrub is necessary in assessing soil stress.

Soil may be classified by either chemical or physical composition. These two properties of soil are both important to plant health. Soil must contain the necessary balance of essential minerals for normal root growth to occur. But soil must also allow the easy passage of roots and must be able to retain the minerals essential for plant growth. Obviously, these two properties are closely interrelated, and information about both physical and chemical soil properties is valuable for diagnosis of soil stress. Chemical stress in soil can be due to unfavorable pH and/or imbalances in nutrients, but it can also be caused by a toxic material in the soil, such as that produced by the roots of black walnut and butternut. Physical stress in soil usually results from factors that restrict root growth. Root girdling is a common result of impervious layers or barriers to root growth.

DEGREE OF ACIDITY OR ALKALINITY (pH)

One important measure of soil condition is the degree of acidity or alkalinity, which is measured in units from 1 to 14 and termed *pH*. Levels of pH from 1 to 7 are called acidic and levels greater than 7 are basic. Most trees grow best in soil pH 5.5 to 6.5 (slightly acid). Lawns grow best around pH 6.5, a little higher than the optimum for most trees.

Acid Soil

Soil with pH below 5 is termed acid soil and is detrimental to the growth of most trees. Many essential soil nutrients are not available to the roots in sufficient quantity at this low a pH, while others may become more available and reach toxic proportions. Acid soil can be corrected by the application of ground limestone to raise the pH to the proper level.

Alkaline Soil

Soil above pH 7 is termed alkaline soil and is also detrimental to the growth of most trees and especially to acid-loving shrubs such as azalea and rhododendron. Alkaline soil also has a limiting effect on the availability of certain soil nutrients. Alkaline soil can be corrected by adding sulfur, aluminum sulfate, or peat to lower the soil pH. Neither acid nor alkaline soil conditions can be accurately diagnosed without the aid of a soil test.

NUTRIENT DEFICIENCIES AND TOXICITIES

There are many inorganic nutrients needed in proper balance for any plant to grow normally. Fortunately, most soils already contain an adequate balance of these essential nutrients; however, there are common instances where a better balance is needed to prevent nutrient deficiency. Essential soil nutrients fall into two major categories: (1) macroelements, and (2) microelements. Macroelements are the most familiar group and include all the major components of fertilizers: nitrogen (N), phosphorus (P), and potassium (K) or potash. The labeling on the fertilizer indicates the ratio of these three macroelements in the order N–P–K. Sulfur (S), magnesium (Mg), and calcium (Ca) are also considered macroelements. However, macroelement imbalance is most commonly associated with N, P, and K. Nitrogen, due to the high solubility of most of its chemical forms, is especially subject to leaching. There are several microelements essential for plant growth, including manganese (Mn), iron (Fe), boron (B), copper (Cu), zinc (Zn), and molybdenum (Mo). Microelement imbalances are generally much less common than macroelement imbalances.

Diagnosis of both macro- and microelement imbalance is difficult. Symptom development is often unreliable because most nutrient disturbances cause similar effects on trees. One of the most useful aids in the diagnosis of nutrient imbalance is a soil analysis. Ideally a soil test should determine both chemical and physical properties of the soil sample. While chemical analysis reveals only the concentration of nutrients in the soil, the examination of the physical condition (amounts of silt, sand, and clay and a soil profile) can reveal the availability of nutrients for root growth. In some cases nutrients may occur in a chemical form unavailable for root absorption. Results from a soil test may indicate a normal concentration of soil nutrients but may not show that very little of the essential nutrient is being absorbed by the roots. Therefore, chemical analysis of foliage is often required in addition to a soil test to achieve an accurate indication of the nutrient condition of the tree. Careful inspection of the soil and roots is also required, and inquiries about recent activities in the soil around the tree should also be made before a nutrient imbalance is suspected.

Soil Testing

Most state universities have the facilities to analyze soil samples for pH and macroelement content. Many institutions still analyze the soil for concentrations of silt, sand, and clay. Some can even analyze for microelement content. This service is usually free or available at a nominal fee to the public. Several samples of the top 12 inches (30 cm) of soil should be collected over a representative area around the affected tree. The individual samples should then be mixed together. The combined sample should first be clearly labeled as to the type of tree already growing in the soil or the type of tree to be planted in the soil. This information, along with the person's name and address, should be brought or mailed to the nearest testing facility as soon as possible.

Foliar Analysis

A few large universities and many private testing laboratories can analyze leaves and other plant tissues for a large number of macro- and microelements. Most commercial fruit growers analyze their trees at least once every few years. A fee is usually charged for this service. Leaves from an affected tree or shrub should be collected and air dried. The dried leaves are then sent to the testing facility as soon as possible.

Specific Diseases Caused by Nutrient Imbalance

A large number of disorders resulting from nutrient imbalance have been reported. However, only a few have occurred in sufficient frequency to be considered problems to the arborist. The following disorders, therefore, will be discussed in detail: iron deficiency chlorosis, copper toxicity, boron toxicity, and manganese deficiency.

Iron Deficiency Chlorosis   Iron, in alkaline soils, is often in a form too tightly bound to other molecules to be available for root growth. This problem occurs most frequently in high calcium soils and is sometimes termed "lime-induced" chlorosis. Trees and shrubs growing on these soils often suffer from an iron deficiency. Plants in this condition can no longer produce the chlorophyll that is essential for the manufacture of energy reserves via photosynthesis.

A general yellowing of most leaves is the first visible symptom of iron deficiency chlorosis. Closer inspection will reveal the yellowing to be most intense between the veins and at the leaf margins. Any remaining normal green leaf tissue will occur along the veins. The chlorotic condition may appear in a range from yellow-green to lemon-yellow to almost white depending on the severity of the iron deficiency. Early in the disease progression leaves may flush appearing normal both in color and size, but new leaves produced at the shoot tips become increasingly chlorotic as the growing season continues. Later in the progression chlorotic leaves emerge in the spring and steadily decrease in size. Shoot growth is also stunted. Interveinal necrosis can also occur, and this condition may often be confused with scorch from water deficiency stress. The effect of iron deficiency on food production and storage after two growing seasons may result in dieback of twigs to progressively larger branches in the upper crown. Sprouting or suckers consisting of chlorotic foliage may appear on the trunk and larger branches. Decline symptoms continue to increase unless therapeutic treatments are given. Without treatment, death usually occurs in a few years.

A large number of trees and shrubs are susceptible to iron deficiency chlorosis (Table 17.1). This problem can be avoided by selecting resistant species in high pH calcareous soils. Iron deficiency chlorosis can be treated by altering the soil environment to correct the nutrient imbalance, or by adding iron therapeutically to the affected tree.

Iron can be applied to the leaves or added to the soil around the tree as iron sulfate or as one of the synthetic iron chelates. Both these forms of iron are soluble in the soil and can be absorbed by the roots. Iron chelates are organic compounds that are bound to iron molecules to prevent the iron from binding with chemicals in the soil such as phosphorus. The chelates themselves are absorbed by the leaves or roots. Iron is applied to foliage in a spray when the leaves are fully expanded. Spraying should be avoided during the blossom period of flowering trees. Iron may also be applied to soil by adding it to holes 6 inches (15 cm) deep evenly placed under the tree, by spreading it evenly over the soil in the same location in a granular form followed by watering or by adding it in a spray.

Both foliar and soil applications of iron salts and iron chelates have resulted in only limited success in the treatment of iron deficiency chlorosis. Iron chelates, however, have proven much more effective. Long-range control of iron deficien-

TABLE 17.1
Some Woody Plants Susceptible to Iron Deficiency Chlorosis

Trees	Trees	Shrubs
American elm	Oak, black	Azalea
American holly	Oak, mossy cup	Forsythia
Bald cypress	Oak, pin	Hydrangea
Birch, canoe	Oak, red	Magnolia
Birch, yellow	Oak, swamp white	Rhododendron
Cherry, black	Oak, white	Rose
Cherry, mazzard	Oak, willow	
Cottonwood	Pine, jack	
Eucalyptus	Pine, ponderosa	
Flowering dogwood	Pine, white	
Horse chestnut	Sweet gum	
London plane	Walnut	
Maple, Norway		
Maple, red		
Maple, silver		
Maple, sugar		

cy chlorosis in trees would involve permanent changes in soil pH. Alteration of pH around a tree is extremely difficult because of the large mass of soil involved and the large capacity of soil to resist or buffer against a pH change. To bypass the problem of soil uptake by roots, iron has been injected directly into the trunk.

Placement of iron salts or chelates directly into trees has been used successfully for the treatment of iron chlorosis. Treatments for trunk injection or trunk implantation of iron compounds into trees are commercially available. Wounds are first made approximately every 6 inches (15 cm) around the base of the tree; injection capsules or implants are then inserted into the wounds. The iron-containing materials are taken into the xylem and translocated via the sap stream to the leaves. The iron treatment usually corrects the chlorotic condition in 2 to 4 weeks. Even distribution of the iron materials is critical for the success of the treatment. Uneven distribution of the iron treatment will result in alternating dark green and light green sections of the tree. It is important, therefore, to make the wounds as low on the buttress as possible to ensure the maximum distribution of the iron to all the foliage of the tree. Some conifers, especially pines, must be injected at approximately 5 feet (1.5 meters) aboveground, due to a heavy resin response to wounding in the buttress that may block any movement of materials into the tree (see Chapter 21, Tree Maintenance). In any case, the injected iron will gradually become exhausted in the foliage in two or three years, chlorosis will start to return, and retreatment will be required.

Copper Toxicity   Woody plants can be exposed to toxic levels of copper from air pollution, copper fungicides, and copper-treated burlap. Repeated application of Bordeaux mixture and other copper fungicides can cause a soil buildup of copper that can eventually be toxic to plants. If the burlap used for balled-and-burlapped stock has been treated with copper sulfate, the enclosed plants may be injured from the copper that leaches into the soil ball. Runoff from copper leaders and drainpipes has also been considered toxic to trees and shrubs around homes. Copper sulfate applied to ponds and lakes to control algae can be toxic to trees growing at the shore when the chemical is used in high concentrations.

Symptoms of copper toxicity are similar to many other nutrient disorders, such as iron deficiency chlorosis, and also to other noninfectious diseases. Leaves display interveinal chlorosis and the plants are generally stunted. Root growth is sparse, with thicker main roots but few laterals. Necrotic lesions may also appear. Plants will continue to decline at a rate dependent on copper concentration and eventually will die. Occasionally copper toxicity is misdiagnosed as iron deficiency chlorosis, but plants do not respond completely to the iron treatments because of the root injury caused by the copper.

Soil analysis usually does not reveal most copper toxicity problems because they occur at such low levels. Kuhns and Sydnor (1976) found soil concentrations of copper from 1 to 50 ppm caused chronic symptoms on azalea, boxwood, and cotoneaster. The toxicity of copper is also dependent on physical soil properties, with toxicity being least in light sandy soil and progressively higher in heavier soils with a high organic matter or clay content. Tissue analysis is the most reliable diagnostic measurement of copper toxicity. Although roots, stems, or leaves may be used for analysis, the concentration of copper in root tissue is the best indicator of disease severity.

Copper toxicity can be treated most successfully by removing the source of copper from the trees. Copper sulfate-treated burlap should be avoided for balled-and-burlapped stock. Runoff from copper drainage systems should be directed away from trees and shrubs and into a sewer or dry well. The concentration of copper sulfate in ponds and lakes should be kept below 1 ppm, which should control algae and not injure trees. Copper from air pollutants is difficult to control because the source may be far removed from the affected plant. Control of air pollution usually requires a community-wide effort involving several levels of government (both state and federal) to stop emissions at the source (see Chapter 22, Air Pollution). Therapy applied after removing the source of copper involves a program of fertilization using high phosphate fertilizers for root growth, and includes watering.

Boron Toxicity Boron is an essential microelement that may cause injury to plants when soil concentrations are too high. High concentrations of boron are found in borax detergents. If waste water containing such detergents is regularly flushed on soil around trees and shrubs, boron toxicity may result.

Boron toxicity in pines has resulted in progressive tip necrosis. In severe cases needles range from green at the base to yellow and then to brown at the tip. The needles eventually die and are cast. Decline and death result if the source of toxicity is not removed. Boron toxicity can be diagnosed most accurately with analysis of foliage, although soil analysis may also be helpful.

Boron toxicity can be treated, like copper toxicity, if additional boron is not added to the soil. Waste water from laundries using borax detergents should be added directly to the sewer system or into a dry well. Normal soil leaching will lower boron levels to nontoxic levels in 1 or 2 years in most cases. Watering and fertilization will stimulate new growth and also aid in boron leaching.

Manganese Deficiency   Manganese in high pH soils is often unavailable to roots and a deficiency may occur. This problem is most common on maples, and has been termed maple chlorosis because of its similarity to iron deficiency chlorosis, which also occurs on alkaline soils.

The symptoms of manganese deficiency are similar to those of iron deficiency, namely, foliar chlorosis, both marginal and interveinal, and twig dieback in the upper crown. Trees appear progressively more chlorotic over the growing season and exhibit early fall coloration. Trees may eventually decline and die if not treated. Soil analysis may indicate a normal level of manganese but at high pH levels the manganese cannot be absorbed. Foliar analysis is required to accurately diagnose manganese deficiency.

Manganese deficiency is treated by applying manganese sulfate or a manganese chelate to the foliage or directly into the tree. Two sprays are applied to the fully emerged foliage each season. Trunk injection is applied via capsules in a manner similar to iron treatments. In experiments by Kielbaso and Ottman (1976), recovery of maples was slower from trunk-injected manganese than that commonly reported for iron deficiency chlorosis.

Manganese deficiency is also a common problem in some palms, especially in winter when colder soil temperatures inhibit the roots' ability to absorb that element from soils. Affected palms develop terminal and marginal necrosis on their fronds, especially near the top or growing point of the plant. The condition is often referred to as "frizzle top" of palms.

PHYSICAL CONDITION OF SOIL

Soils, regardless of nutrient content, must not restrict root growth if trees are to survive. Soil around homes and along streets is often in poor physical condition. Although the original soil in most locations is adequate for plant growth, changes have usually occurred in the soil structure during road and house construction. Soils where shade trees are often planted may be low in organic content, compacted, or otherwise altered from the well-drained loam that is ideal for the growth of most woody plants.

Newly planted trees should, of course, be placed in good quality soil but soils around existing trees in many cases should also be improved to allow increased root growth. Organic matter is one of the most common components that is deficient in soil around trees. If soils have experienced heavy foot traffic, a compact high-clay soil may result. This can be corrected by first aerating the soil with a crow-bar or soil needle and amending the soil with sand, vermiculite, perlite, or organic materials, such as bark mulch, or well-rotted manure. These materials may be mixed to achieve a light, well-drained soil high in organic matter. Whenever organic amendments, except manures, are added it is recommended that a small amount of fertilizer (1 lb/1000 ft^2) be added to balance any loss of nitrogen that may occur during decomposition of the organic matter.

Impervious layers of clay or hardpans may occur in the subsoil and, acting as dams for surface water, prevent adequate drainage away from roots. These layers must be broken through, usually with hand digging tools and occasionally with larger equipment, or saturated soil with subsequent root suffocation can occur. Testing for hardpans is a recommended procedure before planting any tree or shrub (see Chapter 21, Tree Maintenance).

ROOT GIRDLING

Root systems of trees and shrubs normally grow outward in a radius from the trunk. Occasionally one or several roots may begin to grow back across part of the main root system (Fig. 17.1). After several years the turned root(s) will "strangle" or "girdle" the other roots, thereby preventing adequate water and nutrient flow up the tree.

Girdling roots are most often associated with some type of restricted root space. The restrictions may be a street, sidewalk, driveway, building, well, or any other obstacle that impedes the normal outward growth of roots (Fig. 17.2). A large boulder, ledge, or compacted soil may also contribute to the occurrence of girdling roots. Root injury, such as commonly associated with trenching, can cause the formation of new roots close to the trunk, which often curve around and form girdling roots.

Root girdling, like most root problems, is often hard to diagnose without close inspection since the girdling root is usually located below the soil line. The aboveground symptoms often resemble those of early tree decline. Leaves may be smaller than normal and may show symptoms of scorch. A large amount of seed may be produced. In progressive cases of root girdling small twigs and increasingly large branches may die in the upper crown. The lower leaves and branches, in general, remain healthy; however, symptoms may appear first in a lower branch directly above a girdling root. If the girdling root is near the surface it may have restricted the normal buttress swell of the tree, making it appear to enter the ground like a telephone pole on one side (Fig. 17.3). Deep planting and fill over roots, however, can also result in similar appearance (see Chapter 19, Construction Injury and Soil Competition). The lack of a complete root flare or, in some severe cases, an indentation of the trunk into the soil are reliable indicators of girdling roots below. In many cases, however, the root flare appears

Fig. 17.1   Root girdling on a sugar maple. (Photo courtesy of Shade Tree Laboratories, University of Massachusetts, Amherst.)

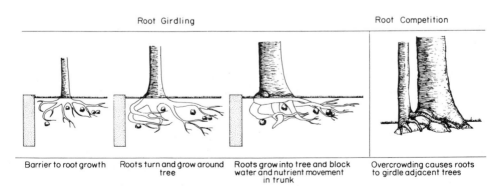

Fig. 17.2   Root girdling and root competition.

Fig. 17.3 Girdling root on a sugar maple just below the ground level. Note the lack of roof flare and slight indentation of the trunk at the soil line. (Photo courtesy of Shade Tree Laboratories, University of Massachusetts, Amherst.)

normal and only careful excavation of the entire root system close to the tree to a depth of at least 18 inches (45 cm) will reveal the presence of a girdling root.

Root girdling is best controlled by avoiding locations where there is limited root space. Fast-growing trees are best reserved for open areas while smaller or dwarf trees will grow best in confined locations. Root girdling can be treated successfully in its early stages by surgical excision of girdling roots. The root is severed in two places to allow normal growth of the constricted portion. In advanced cases, where several large girdling roots have encircled most of the root system, it is doubtful that attempts at surgical removal will be successful. In these cases the girdling roots may function significantly in root absorption and thus their removal would be harmful to the tree. Removal of dead and dying branches and general crown thinning are recommended to slow the rate of decline.

ROOT COMPETITION

Trees growing in dense groups compete for available root space. Roots of one tree occasionally cross the buttress flare of another and grow into it (Fig. 17.2). The effect is the same as previously described for girdling roots, except in this case the root originates from another tree. Root competition can be prevented by proper spacing of plantings or by thinning excess trees in natural stands. Treatment is also by the surgical excision method used for girdling roots.

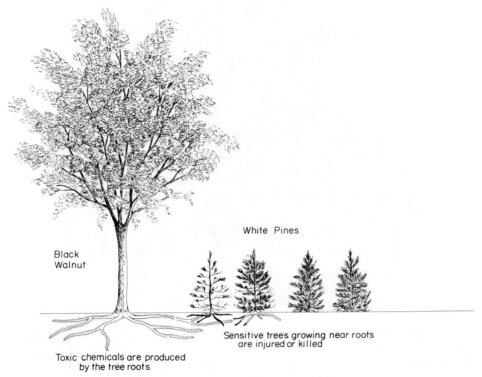

Black
Walnut

White Pines

Sensitive trees growing near roots
are injured or killed

Toxic chemicals are produced
by the tree roots

Fig. 17.4   Walnut or butternut toxicity.

JUGLANS TOXICITY

Black walnut and butternut both release a chemical (juglone) from their roots during the growing season that is toxic to many species of plants (Fig. 17.4). It is felt that the production of this "natural herbicide" has evolved in the species to decrease competition from surrounding trees. This phenomenon of plant : plant interaction is known as allelopathy. The toxic effect of juglone on other plants has been called juglans toxicity or walnut wilt.

Trees and shrubs growing near a black walnut or butternut may suddenly wilt or turn chlorotic and often die during the growing season. Trees growing near the roots of walnut and butternut are often killed and vascular discoloration similar to that produced in wilt diseases may also occur in the main stem. Apple, mountain laurel, pear, pine, rhododendron, and sour cherry are some woody plants known to be susceptible to juglone toxicity.

Juglans toxicity can be prevented by not planting sensitive plant species under the dripline of black walnut and butternut trees. Sensitive species planted beyond the root zone will not be affected. Another control is the removal of the black walnut or butternut tree. Juglone does not persist in the soil over winter; therefore, sensitive species may be planted in the same location a year after a butternut or black walnut is removed.

LIGHTNING INJURY

Electrical discharges (lightning) during storms occasionally strike shade trees (Fig. 17.5). Those in exposed locations such as open fields or on hilltops or trees above the forest canopy are most commonly struck. Injury from lightning is variable, ranging from the explosion or burning of the entire tree to minimal damage to the trunk and roots. In many cases, however, when only minor injury is evident on the trunk, considerable injury has occurred to the roots (Fig.

Fig. 17.5 Lightning injury on white oak. Note band of bark killing from the soil line up the tree. (Photo courtesy of Shade Tree Laboratories, University of Massachusetts, Amherst.)

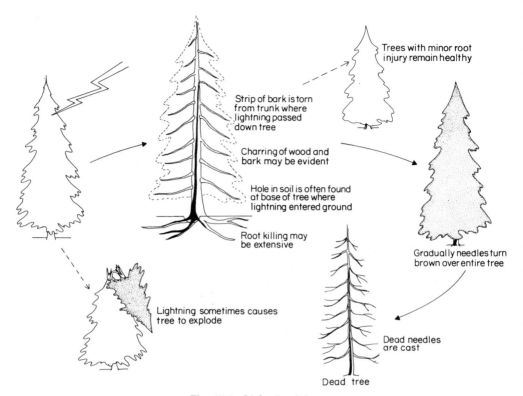

Trees with minor root injury remain healthy

Strip of bark is torn from trunk where lightning passed down tree

Charring of wood and bark may be evident

Hole in soil is often found at base of tree where lightning entered ground

Root killing may be extensive

Gradually needles turn brown over entire tree

Lightning sometimes causes tree to explode

Dead needles are cast

Dead tree

Fig. 17.6 Lightning injury.

17.6). The roots of a tree that has been struck with lightning should be inspected for extent of injury before therapeutic treatments begin. If more than 50% of the root system appears healthy, prompt therapy may help the tree to recover. Trees injured by lightning should be treated similarly to other stressed trees. Loose bark should be removed and injured bark cut back to healthy tissue. The injured tree should be fertilized with compounds low in nitrogen but high in phosphorus and watered during dry spells. Frequently, trees may be subject to repeated strikes due to their exposed location and in these cases installation of lightning protection is recommended. Lightning protection for trees is similar to the lightning rods that protect buildings. A copper cable is placed as high as possible in the tree and fastened, with copper nails, to the trunk all the way down the tree to the ground. The end of the cable is attached to a ground rod driven deep into the soil. This system, like all devices attached to living trees, should be checked periodically and adjusted to allow for growth and expansion of the tree.

LITERATURE CITED

Kielbaso, J. J., and K. Ottman. (1976). Manganese deficiency—contributory to maple decline. *J. Arbori.* **2**, 27–32.

Kuhns, L. J., and T. D. Sydnor. (1976). Copper toxicity in woody ornamentals. *J. Arbori.* **2**, 68–72.

SUGGESTED REFERENCES

Brooks, M. G. (1951). Effect of black walnut trees and their products on other vegetation. *W. Va., Agric. Exp. Stn., Bull.* **347**, 1–31.

Kuhns, L. J., and T. D. Sydnor. (1975). The effects of copper treated burlap on balled and burlapped *Cotoneaster divaricala*, Rehd and Wils. *HortScience* **10**, 613–614.

Messenger, A. S. (1976). Root competition: grass effects on trees. *J. Arbori.* **2**, 228–230.

Neely, D. (1976). Iron deficiency chlorosis of shade trees. *J. Arbori.* **2**, 128–130.

Neely, D., and D. F. Schoeneweiss. (1974). Correction of pin oak chlorosis. *Arborist's News* **39**, 37–40.

Plass, W. T. (1975). An evaluation of trees and shrubs for planting surface-mine spoils. *U.S., For. Serv., Res. Pap.* **NE-317**, 1–8.

Sinclair, W. A., and E. L. Stone. (1974). Boron toxicity to pines subject to home laundry waste water. *Arborist's News* **39**, 71–72.

Stone, E. L., and G. Baird. (1956). Boron level and boron toxicity in red and white pine. *J. For.* **54**, 11–12.

Van Camp, J. C. (1961). Tolerance of trees to soil environment. *Proc. Int. Shade Tree Conf.* **37**, 5–19.

18

Animal Injury

INTRODUCTION

Animals (not including insects, mites, or nematodes) can injure trees in three general ways. They can (1) wound trunks, branches, and roots; (2) place toxic chemical products on tree tissues and the soil; and (3) cause soil compaction. Most wounds are caused by animals chewing, pecking, clawing, or digging on or around trees (Fig. 18.1). Low foliage and shoots can be chemically poisoned

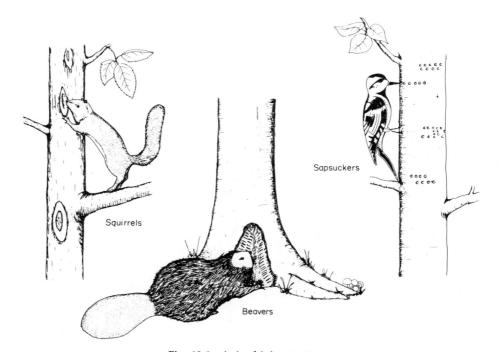

Squirrels

Beavers

Sapsuckers

Fig. 18.1 Animal injury to trees.

by urine. The soil environment around trees is usually altered by chemical waste products or by the compressive effect of animal feet.

ANIMAL WOUNDS

Chewing Injury

Wounds from chewing are the most common type of animal injury on woody plants and are most frequently associated with rodents such as mice, rabbits, squirrels, porcupines, and beaver (Fig. 18.2). However, deer, cattle, horses, moose, elk, and other grazing animals also chew trees (Fig. 18.3). Most of these

Fig. 18.2 Injury on beech from beavers. (Photo courtesy of Alex L. Shigo, USDA Forest Service, Durham, New Hampshire.)

Fig. 18.3 Injury to aspen from elk. (A) Recent damage above and old healed wounds below. (B) Tree at right has new and old wounds, tree at left has canker that has followed the chewing wounds. (Photo courtesy of USDA Forest Service.)

animals are seeking food from tender twigs and green bark, but some simply chew bark out of curiosity or boredom (Fig. 18.4). Most chewing injury occurs in the winter when other sources of food are not available. Rodents are the most destructive group because they often feed on young seedlings in nurseries or on newly planted trees. Rodents often completely remove the bark of seedlings and young trees beneath the snow and can cause considerable losses in tree nurseries and plantations (Fig. 18.5). Chewing injury can be controlled to some extent with animal repellants, such as thiram, applied in the fall. In areas where high populations of rodents are common, hardware cloth fencing should be placed from the soil surface to above the snow line around each tree. Population control with poison bait can also be used in some cases. The use of many poison baits, however, is now restricted and homeowners are strongly advised to check local pesticide regulations before using any poison baits. Since most chewing animals are closely related to animal pets and humans, improperly used poison baits can be a hazard to the entire community.

Pecking Injury

Pecking wounds from birds can also be a problem on trees. The yellow-bellied sapsucker is thought to feed upon sap; however, some evidence indicates that

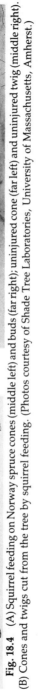

Fig. 18.4 (A) Squirrel feeding on Norway spruce cones (middle left) and buds (far right); uninjured cone (far left) and uninjured twig (middle right). (B) Cones and twigs cut from the tree by squirrel feeding. (Photos courtesy of Shade Tree Laboratories, University of Massachusetts, Amherst.)

Fig. 18.5 (A) Mouse damage to aspen bark beneath the snow. (B) Close-up. (Photo courtesy of USDA Forest Service.)

the sapsucker feeds on insects attracted to sap. This bird pecks even rows of holes around the tree and is perhaps the worst offender (Fig. 18.6). These pecking wounds, when numerous, result in cankers that can kill patches of bark and, if sufficiently severe, can girdle the top of the tree (Fig. 18.7). The yellow-bellied sapsucker has a wide tree host range and some of the preferred trees are thin-barked: apple, birch, Chinese elm, hemlock, maple, and Scots pine. The only known control, short of removing the sapsuckers, is to remove any potential nesting tree (large poplar trees with internal decay) in the area.

The pine grosbeak is another tree pest, which feeds on terminals of young conifers and can be a severe problem on conifer nurseries and Christmas tree plantations. Sapsuckers and pine grosbeaks should not be confused with woodpeckers that are only seeking insect larvae in already dead and decayed wood in trees. For the most part woodpeckers are serving a beneficial function by alerting the homeowner to decay in trees and should not be disturbed.

Miscellaneous Animal Wounds

Confined farm animals and pets can also cause considerable injury to trees (Fig. 18.8). Cattle and horses often remove trunk and root bark with their hooves, especially in rain softened soil or in marshy areas. The root areas of trees

Fig. 18.6 Sapsucker injury on hemlock. Note internal defects caused by old injury in split section at left. (Photo courtesy of Alex L. Shigo, USDA Forest Service, Durham, New Hampshire.)

adjacent to or inside hooved-animal enclosures should be fenced off to protect them.

Cats often drag their claws over bark of trees causing great bewilderment to homeowners as to the nature of the shredded bark. Dogs of all ages will often chew on bark of trees and shrubs when they are confined. They will also dig in soil around plants and cause injury to roots. Pet repellants can protect most high-value trees and shrubs from pet injury.

Treatment of Animal Wounds

Animal wounds should be treated like other mechanical injuries (see Chapter 13, Wound Diseases). Torn or loose bark should be removed and bark cut back until living tissues are found all around the wound. The wound should be shaped in a vertical ellipse. Bark around old wounds should be removed until

Fig. 18.7 Sapsucker injury on (A) sugar maple; (B) white birch. Note extensive bark killing beneath the injury on sugar maple and the trunk swelling from repeated injury on white birch. (Photos courtesy of Alex L. Shigo, USDA Forest Service, Durham, New Hampshire.)

callus is exposed. In extreme cases of girdling, twigs can be grafted across the wound ("bridge grafting").

TOXIC CHEMICALS ON TREES AND SOIL

Toxicity from urine is perhaps the most troublesome form of soil environment alteration by animals. Dogs are the most numerous and worst offenders around shade trees and shrubs. Male dogs use trees and shrubs for territorial markings and often the same plant is urinated upon by every passing male dog. Urine is a strong alkaline chemical that is damaging to foliage and branches. Injury is often most severe on evergreens. The most common symptoms are browning of the low branches (below 2 feet) in one location, usually near steps or at corners of buildings. The urine of female dogs often causes brown patches on lawns and could affect the roots of shrubs and small trees nearby. Although most injury is due to direct urine spray of foliage, there is some feeling that in high-use urban areas a chemical modification of the soil may result from animal urine and feces. If included in animal enclosures, such as chicken yards, trees sometimes die

Fig. 18.8 Severe bark injury to ash from chewing by horses.

from the intensity of the manure. Fences and pet repellants can often prevent animal injury on preferred shrubs and trees.

SOIL COMPACTION

Soil compaction by chained or penned animals can also harm trees and shrubs. Soil physically compressed in this manner is less permeable to moisture, nutrients, and gases. Most soil compaction, however, is caused by humans (foot and vehicle traffic), so it is detailed under people-pressure diseases (Chapter 19, Construction Injury and Soil Compaction). Since animal and human soil compaction is the same, control recommendations are the same. Mulch should be put around trees and shrubs most likely to be affected *before* the soil compaction occurs. Already compacted soil must be loosened (an air hose attached to a root-feeding needle, or a crowbar, works quite well) and then mulch applied.

SUGGESTED REFERENCES

Dimock, E. J., II, R. R. Silen, and V. E. Allen. (1976). Genetic resistance in Douglas-fir to damage by snowshoe hare and black-tailed deer. *For. Sci.* **22,** 106–121.

Krebill, R. G. (1972). Mortality of aspen on the Gros Ventre elk winter range. *U.S., For. Serv., Res. Pap.* **INT-129,** 1–16.

Ostry, M. E., and T. H. Nicholls. (1976). How to identify and control sapsucker injury on trees. *U.S., For. Serv., North Cent. Exp. Stn.,* pp. 1–5.

Rushmore, F. M. (1969). Sapsucker damage varies with tree species and seasons. *U.S., For. Serv., Res. Pap. NE* **NE 136,** 1–19.

Shigo, A. L. (1963). Ring shake associated with sapsucker injury. *U.S., For. Serv., Res. Pap. NE* **NE 8,** 1–10.

Smith, E. M. (1975). Mice are choosy—like euonymus best! *Ohio State Co-op Ext. Serv., Nursery Notes* **8,** 5–6.

19

Construction Injury and Soil Compaction

INTRODUCTION

Trees, like people, are easily disturbed by changes in their surroundings. Building and road construction is one of the most damaging forms of people-pressure disease. The detrimental effects of people's activities on soil structure also cause considerable losses of shade trees. However, it may take several years before these stresses result in obvious symptoms. Diagnosis may be difficult because symptom expression is not specific for these stresses. Major emphasis, therefore, will be upon the types of activities resulting in construction injury and soil compaction and upon their prevention and treatment.

CONSTRUCTION INJURY

Construction activities cause a complex series of site changes that often lead to the decline and eventual death of established trees. Most injuries result from (1) change of grade and (2) mechanical injury to the trunk and roots.

Change of Grade

In most construction projects, land around new buildings or roads is graded level or into a gentle slope. Since the natural ground level to which established trees have adjusted the growth of their root system is usually uneven, grade changes around trees commonly occur (Fig. 19.1). To achieve the desired site preparation the builder may either raise or lower the grade around trees. These changes in the original grade cause considerable injury and often the eventual death of the affected trees.

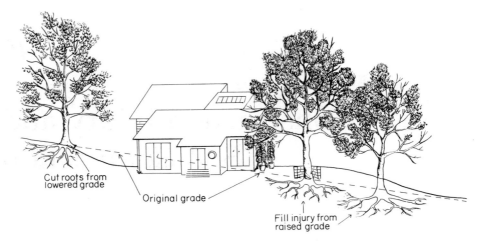

Fig. 19.1 Change of grade around a new house.

Raise of Grade Raising the grade or filling over the roots of trees often results in suffocation of the roots. Roots absorb oxygen and give off carbon dioxide during respiration, and any impervious layer over them will cause anaerobic conditions that will be detrimental to roots. The detrimental effects of the fill are directly related to its depth. The physical composition of the fill also contributes to its effect on trees. Sand or gravel fills are much less severe than heavy clay soils.

Impervious coverings such as asphalt and concrete placed over the roots of an existing tree may also cause a suffocation stress similar to fill injury. This should not be confused with trees planted in narrow soil space such as along city streets. In the latter case young planted trees are able to adapt to the soil conditions by growing slowly and seeking the most favorable areas in which to extend their roots. Roots of established trees, in most cases, cannot tolerate these changes of the soil conditions.

Some species may tolerate fill over roots better than others, but all are injured by raised grade. Sometimes fill is temporarily placed around trees during construction and later removed. If the fill is over the roots during the growing season, suffocation stress most likely would cause injury to roots. The longer the fill remains over the roots, the greater the injury. The effects, however, may not appear as symptoms for several years after the initial stress.

Small wells are often placed around filled trees in an attempt to alleviate the effects of the raised grade (Fig. 19.2). Although this practice may appear aesthetically pleasing, it does not correct the major problem: fill over most of the root system. Roots of most trees extend to the drip line and often beyond. A small well around the trunk will not uncover enough of the root system to prevent fill injury. A well must extend to the drip line to prevent the effects of fill injury.

Fig. 19.2 (A) Raising of grade around a sugar maple. Note thin foliage indicative of early stages of decline. (B) Close-up showing trunk in a small well. Note extensive bark killing on trunk. (Photos courtesy of Shade Tree Laboratories, University of Massachusetts, Amherst.)

This is, of course, difficult and often impossible to achieve in many locations. The chances of saving a filled tree are usually proportional to the size of the well placed around it. Wells placed around established trees with raised grades should not be confused with planting trees in wells. Young planted trees, whose roots are initially contained within the well, can adapt to the soil level around them as they grow into larger trees.

The effects of fill are not distinctive and can be easily confused with a variety of infectious and noninfectious root stresses. Among the earliest symptoms are small leaf size and premature fall coloration. Dieback of twigs and progressively larger branches occurs in the upper crown. Sucker growth may occur on large branches and on the trunk. Decline progresses from top to bottom and the tree eventually dies. Death may occur as a direct result of the fill or may involve a weak parasite that can attack trees primarily when they are in decline.

Fill injury should be suspected whenever the contours of the land are altered during the construction of a home or a highway. Fill injury is also common around any areas of recent construction. One of the best ways to detect fill injury is to determine at what depth the normal root flare or buttress swell occurs. If the tree enters the soil without any widening, similar to a telephone pole, there is good evidence for fill over roots (Fig. 19.3). Root flares are sometimes difficult to detect in small trees or in trees with abundant uncut turf at the base. A more

Fig. 19.3 Raising of grade around a sugar maple caused it to enter the ground without a normal root flare. (Photo courtesy of Shade Tree Laboratories, University of Massachusetts, Amherst.)

accurate estimate of the extent of fill must be determined by excavation around the trunk with a hand trowel.

Fill injury is difficult to treat effectively because it is usually not diagnosed until severe or chronic decline symptoms have occurred. Unless fill is removed before a growing season has passed, considerable injury to the root system will have already occurred and cannot be completely reversed by simply removing the fill. Occasionally trees survive fills by the production of a secondary root system. It arises from the trunk just below the new grade level. This new root system gradually takes over the necessary function of the original root system that was killed. Duling (1969) reported that if cuts were made around the trunk just below the new ground level, most trees produced secondary root systems and survived fill injury. Unfortunately little additional information exists on the success of this technique for treating fill injury.

Trees to be filled may be protected by a system of gravel and aerating and drainage tiles around the root system (Fowler *et al.*, 1945). This type of system, when installed *before* any disturbances in the grade around trees occur, can provide aeration of the roots under the added fill and allow a tree to survive where it would almost certainly have died otherwise. It must be emphasized that unless extensive root disturbances by heavy earth-moving equipment over the root systems are prevented during the construction process (see the section on prevention of construction injury, in this chapter), sufficient damage may occur to the root system to prevent an aeration system from saving a tree.

Aeration systems have not been used extensively due to the expense of installation. Recently, however, with use of new construction and landscape materials, such as porous landscape fabrics and perforated plastic pipe, installation of aeration systems on trees of high value has become more common.

If severe decline symptoms are evident and the depth of fill is in excess of 12 inches (30 cm) and extends over the entire root system, the tree should be removed. If decline symptoms are mild or absent or fill is less than 12 inches, therapeutic treatment can sometimes prolong the life of the tree. The chances of saving a tree increase as the depth of fill and the amount of root area under fill decrease. The physical nature of the fill is often as important as the fill depth. Sandy soils are permeable to both water and gases while heavy clay soils or subsoil can be more detrimental to a tree than a much thicker layer of sandy soil, sand, or gravel.

Therapeutic treatments, where applicable, are designed primarily to restore a balanced root : shoot ratio and stimulate root growth. The crown should be thinned at least 20% on declining trees, including removal of all dead and dying branches. Trees showing mild or no symptoms should be thinned 10% a year for 2 years and watched carefully for at least 10 years. Additional thinning may be necessary if decline symptoms reappear. Application of balanced slow release fertilizer designed for use on trees will help stimulate root growth. Fertilizer may be applied to the soil, in either liquid or solid form, or may be injected directly into the trunk. Fertilizer should be applied over the entire area under the drip line and approximately 3 feet (1 meter) beyond. Special effort should be made to fertilize areas where the roots were not covered with fill or where the fill was shallowest. Often a few surviving roots may take over the function of the entire root system, and if their growth can be stimulated, the chances of saving an affected tree are increased.

Lowering of Grade Most of the root system of woody plants is located in the top 18 inches (45 cm) of soil, and a large amount of the small feeder roots are located in the top 6 inches (15 cm) or less. Removal of soil around trees, no matter how minimal, can lead to extensive root injury (Fig. 19.4). Lowering of grade during construction most often results in complete severing of most of the root system where the grade has been lowered (Fig. 19.5). The severity of injury depends upon how much of the grade under the drip line of the tree has been lowered. Cutting of large roots close to the trunk causes much more injury than cutting smaller roots near the drip line. In some cases where roots have been severed by lowering of grade all around the tree, the lack of roots to anchor the tree has resulted in blowdown.

Cut-away showing root injury below ground

Fig. 19.4 Construction injury from lowering of grade.

Fig. 19.5 Lowering of grade in preparation for sidewalk construction has cut most of the roots on one side of this sugar maple. (Photo courtesy of Shade Tree Laboratories, University of Massachusetts, Amherst.)

The symptoms resulting from lowering of grade are much like those resulting from raising of grade. Progressive decline symptoms associated with root injury usually occur, including early fall coloration, small leaf size, twig and branch dieback, and suckering on the trunk and large branches. Injury from lowering of grade can be detected by locating zones where abrupt changes in the normal grade around trees occur. These zones usually occur at the edge of the area planted in turf. In many cases broken roots can be seen around the root flare and under the branches but often hand excavation around roots is necessary to detect injury, especially where turf is growing around the trees.

Lowering of grade, like raising of grade, is difficult to treat effectively once chronic or severe decline symptoms have occurred. If the injury is detected before symptoms occur and does not involve more than 50% of the root system, corrective measures are often successful. Loss of roots from lowering of grade results in a root : shoot imbalance that severely stresses the affected trees. Without a complete root system there is not enough water and nutrients being supplied to the tree to support the needs of the crown. The upper crown which is most distant from the roots most often suffers first. Therefore, the crown dies back as a result of the disturbance in the root : shoot balance.

The extent of root loss from lowering of grade should first be determined. Damaged roots should be pruned and torn bark cleaned back to living tissue (see Chapter 13, Wound Diseases). The root loss must be balanced by a similar reduction in the crown. Phosphate fertilizer should be applied to encourage root growth, but nitrogen-containing fertilizers should not be used for at least 5 years after therapy treatments begin. Affected trees should also be watered during dry periods to help relieve moisture stress on a damaged root system caused by transpiring leaves.

When affected trees are near buildings or roads, it must be determined whether they pose a hazard. Trees that may be potentially hazardous should be cabled to noninjured trees or removed (see Chapter 26, Living Hazard Trees).

Effects of Grade Changes on Soil Water   In addition to the immediate effects of grade changes on trees there are some important changes that occur to the entire site around the affected trees. When fill is added over the soil, the water table tends to rise to achieve its original depth below the soil surface. Since soil oxygen is very low near the water table, suffocation of roots may occur from the saturated soil as well as from the impervious fill above. When the grade is lowered the water table drops back to its original depth. The effect of cut roots from lowering of grade is often exaggerated on the remaining root system by loss of available soil water.

Mechanical Injury to Trunk and Roots

Earth-moving equipment often causes injury to the trunks and roots of existing trees in addition to changing the grade. This type of mechanical injury takes two general forms: (1) surface grading injury and (2) trenching injury.

Fig. 19.6 Injury to trunk and roots by bulldozers. (Photo courtesy of Alex L. Shigo, USDA Forest Service, Durham, New Hampshire.)

Surface Grading Injury   Surface grading by bulldozers and other mechanical surface graders severs and injures large numbers of roots when surface areas around buildings and roads are leveled (Fig. 19.6). In addition, trunks are often injured or "skinned." In these cases no drastic grade changes occur, but most often surface vegetation (small trees, shrubs, leaf litter, etc.) between trees is scraped from the soil surface. In most cases much of the topsoil, which contains large numbers of feeder roots from adjacent trees, is also removed. The removed soil is often screened, replaced, and seeded for turf around the remaining trees.

Injury from surface grading, like most root injury, is usually hidden by turf or a layer of mulch. Injury on the trunk and buttress area from the tractor blade or treads can often be detected. It can be expected that any surface grading around existing trees in a wooded lot will cause root injury. Movement of any heavy construction equipment, with either treads or tires, around trees will also cause root injury. Grading injury causes similar nondistinctive decline symptoms as described earlier for changes of grade. The most accurate method of detection is to obtain the recent history of construction activities around the affected trees.

Most surface grading does not cause injury as severe as lowering of grade or trenching, and many affected trees survive with minimal crown symptoms. However, this injury and/or the subsequent site changes that follow kill a large number of trees. Loss of the humus layer, the topsoil, and a large amount of the feeder roots is certainly detrimental to the tree. The replacement of a thin layer of topsoil with turf growing on it may only compound the injury. Turf competes with the tree's roots for nutrients and moisture and may prevent the injured tree from producing sufficient new roots to recover. In areas where surface grading has already occurred or cannot be avoided, the area under the drip line should be covered with some type of mulch. These materials are similar to the humus layer or duff found in the forest. Mulches prevent rapid evaporation and make water and nutrients available to the roots and thereby create conditions favorable for new root growth.

Phosphate fertilizer should be applied each year for at least 2 years after injury has occurred. A balanced fertilizer then may be substituted if no decline symptoms have occurred. Crowns should be thinned approximately 10%, and additional thinning may be necessary if decline symptoms occur. Watering of affected trees during dry periods is also recommended to partially relieve moisture stress on the injured root system. Any obvious trunk or root injuries should also be properly treated (see Chapter 13, Wound Diseases).

Trenching   Trenching for sewer and water pipes and for underground cables can vertically sever large numbers of roots when the trenches are dug near trees (Fig. 19.7). The amount of injury is directly related to the proximity of the trench to the trunk. Trenches outside the drip line do minimal injury while a trench next to the trunk may cut half of the root system.

Trenching injury is often difficult to detect. The symptom development is similar to that caused by the other forms of construction injury. Broken roots and/or trunk injury occasionally can be detected by careful observation. Fresh

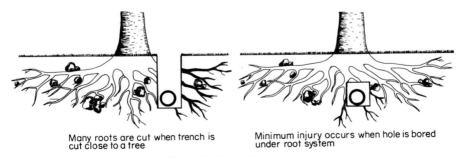

Many roots are cut when trench is cut close to a tree   Minimum injury occurs when hole is bored under root system

Fig. 19.7 Trenching injury.

asphalt or concrete on a street, sidewalk, or driveway that passes in a direct line next to a tree is a clue to recent trenching. Excavation of soil around the tree with a hand trowel is the best way to detect injury and to determine its extent.

Trenching injury, like most other forms of construction injury, is difficult to treat effectively once severe symptoms develop. After severe decline symptoms have occurred, the effectiveness of therapeutic treatments is greatly diminished or nonexistent.

Most injury from trenching around trees can be avoided. Pipes and cables can often be dug in the center of a street to avoid the roots of valuable curbside shade trees. In cases where the trench must pass near the tree most injury can be avoided by a tunnel under the tree's roots. A power-driven soil auger is usually used to dig the tunnel. If trenching is unavoidable, place the trench as far away from the trunk as possible, cut as few roots as possible, and trim all cut roots with pruning tools. All trenches near trees should be refilled as soon as possible to avoid moisture loss around roots, whether or not any root injury has occurred. The crowns of injured trees should be thinned in proportion to the amount of root loss. If the loss of roots is greater than 20%, thinning may be spaced over 2 or 3 years to preserve as much of the original form of the tree as possible. Balanced fertilizer should be applied or injected to stimulate new root growth. Affected trees are drought-prone and should be watered during dry periods to minimize moisture stress.

Injury from Home and Garden Equipment   Trees are often subjected to physical injury from other activities of people that are similar in many ways to construction injury. Mechanical injury also occurs from lawnmowers, snow-plows, automobiles, and a variety of other mechanized equipment that is operated around trees.

Injury is usually caused by contact of this equipment with the trunk and buttress roots. Lawnmowers are the most frequent source of injury since their operation around trees is common (Fig. 19.8). Trees with turf next to trunk and buttress area require close mowing and are often injured by mowers. Mulching around all shade trees would eliminate most mechanical injury from lawn-mowers and any other lawn maintenance equipment. Automobiles, snowplows, and other motor vehicles should be prevented from operating around trees by the erection of barriers such as curbs, large stones, logs, railroad ties, or fences between the road or driveway and the trees (Fig. 19.9).

PREVENTION OF CONSTRUCTION INJURY

Construction injury to trees is often difficult to treat effectively because the injury is too severe for the trees to recover or because the injury is not detected until the trees are in severe decline. However, a large amount of the construction injury around new buildings can be prevented by proper planning before the construction begins. Cooperation between the arborist and contractor is essen-

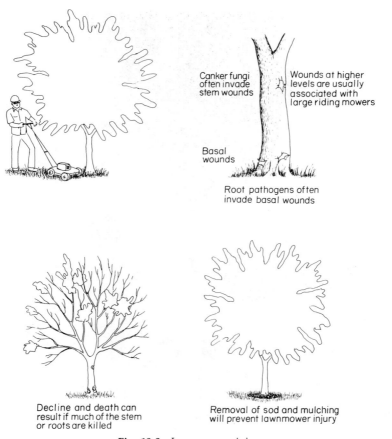

Canker fungi often invade stem wounds

Wounds at higher levels are usually associated with large riding mowers

Basal wounds

Root pathogens often invade basal wounds

Decline and death can result if much of the stem or roots are killed

Removal of sod and mulching will prevent lawnmower injury

Fig. 19.8 Lawnmower injury.

tial. The proposed location on the site of the foundation, driveway, and leach-field should be indicated with stakes by the contractor. Any tree in the staked areas or within 10 feet (3 m) of it should be removed (Fig. 19.10). Trees in the zone just outside the construction area usually suffer severe injury and die within 5 years. A barrier fence should be built all around the 10-foot perimeter bordering the construction area separating it from the trees to be protected. This will prevent earth-moving equipment from encroaching on trees outside the construction area and causing needless injury. The fence should also extend to the side facing the road to prevent movement of earth-moving equipment onto the site except by the driveway. When the construction has been completed the barrier fence can be removed. Although the trees just outside the barrier fence have been protected from most injury during construction they still may suffer from the increased exposure to sun and wind due to the loss of many trees around them. These trees are often subject to sunscald and moisture stress.

Fig. 19.9 Protective asphalt curb around a red maple. Note mulch around base of tree.

Trees on the south-facing edge should be wrapped for sunscald protection. Steps should be taken to keep the trees closest to the areas of construction as vigorous as possible with regular fertilization and watering.

SOIL COMPACTION

Soil structure can be altered by repeated physical compression. Human feet and automobile tires are the most common causes of compression that eventually leads to soil compaction (Fig. 19.11). As the soil becomes more dense it becomes increasingly impervious to surface water and gas exchange. Organic matter decreases and clay content increases. Root growth is severely limited in compacted soil. Most growing roots will turn away from compacted soil just as they turn away from natural barriers in the soil such as boulders. The growth of

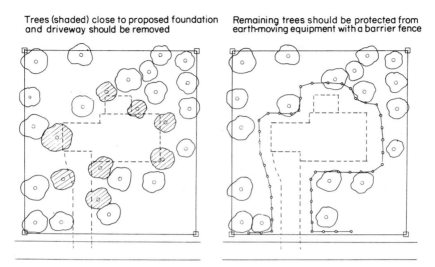

Fig. 19.10 Prevention of construction injury around a new home site.

the trees in or near compacted soil, therefore, will be adversely affected. In mild cases the affected trees may appear completely normal for several years and then only display early fall coloration or minor twig dieback in the crown. In severe cases trees decline and eventually die after several years. Soil compaction injury can be detected by observing areas around trees that have little or no vegetation. These areas are found most often along paths or other high use areas. Compacted soil is difficult to probe with a soil auger or hand trowel and presents a sharp contrast to probing in nonaffected areas containing turf or some other ground cover plants.

The easiest measure of severity of soil compaction is to determine the amount of area under the drip line that is affected. The soil texture, however, also must be considered, since heavy clay soils are more prone to severe compaction than light sandy soils or those high in organic matter.

Compacted soil can be most effectively prevented by keeping foot and automobile traffic away from shade trees (Fig. 19.12). This can be accomplished by erecting barriers to traffic. Curbing in various forms along roads will prevent most cars from driving and parking on the soil around trees. Automobile intrusion around trees is a common problem in parks and other recreation areas where trees line narrow roadways and open campsites. In these locations the perimeter of high use areas may be lined with logs or boulders to contain automobiles in designated parking areas and away from trees. Pedestrians can be encouraged to follow designated walkways by placing fences, park benches, shrubbery, or walls along the edge. A barrier to prevent soil compaction must therefore be constructed to fit the needs of the site and not to detract aesthetically from landscape design.

Fig. 19.11 Soil compaction.

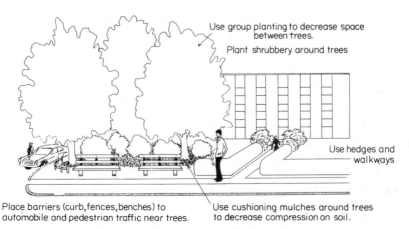

Use group planting to decrease space between trees.

Plant shrubbery around trees

Use hedges and walkways

Place barriers (curb, fences, benches) to automobile and pedestrian traffic near trees.

Use cushioning mulches around trees to decrease compression on soil.

Fig. 19.12 Prevention of soil compaction.

Compacted soil can be treated by loosening the soil to an approximate depth of 6 to 8 inches (15 to 20 cm) with a crowbar or needle-type soil aerator. Organic matter, such as peat moss, or inorganic soil amendments, such as perlite, vermiculite, or sand, should be mixed with the loosened soil to lighten it and increase its resilience. Some type of mulch should be added over the soil to a depth of at least 4 inches (10 cm). There are a large number of mulching materials available (see Chapter 21, Tree Maintenance). Mulches should ideally cover all the area under the drip line, but often this is unreasonable because replanting of turf is desired. In these cases as much as possible of the area under the drip line should be mulched. Mulches should also be amended with approximately 1 lb/1000 ft² (0.45 kg/10 m²) of balanced fertilizer to replace any nutrient loss to mulches. Dead and dying branches should be removed, and an additional 10% crown thinning is also recommended.

Paving blocks or bricks are often considered as an alternative to mulches in the prevention of soil compaction in heavily used sites. Unfortunately, there is very little porosity through these materials, and the roots of trees have been found growing only in the spaces between the blocks and not under them. Recently, porous blocks have been developed for landscape use. It was found, however, that their pores soon became filled with soil once in place, which caused them to become impervious to air movement. The most effective barriers to soil compaction in heavily used urban areas are those that maintain an air gap between the walking surface and the soil over the root system. These systems usually employ raised gratings 4 to 8 inches (10 to 20 cm) above the soil level that can be removed for cleaning and can accommodate growth of the trunk.

LITERATURE CITED

Duling, J. Z. (1969). Recommendations for treatment of soil fills around trees. *Arborist's News* **34**, 1–4.

Fowler, M. E., G. F. Gravatt, and A. R. Thompson. (1945). Reducing damage to trees from construction work. *U.S., Dep. Agric., Farmers' Bull.* **1967**, 1–26.

SUGGESTED REFERENCES

Howe, V. K. (1973). Site changes and root damage. *Proc. Int. Shade Tree Conf.* **49**, 25–28.

Howe, V. K. (1974). Site changes and root damage: some problems with oaks. *Morton Arb. Q.* **10**, 49–53.

Messenger, A. S. (1976). Root competition: grass effects on trees. *J. Arbori.* **2**, 228–230.

O'Rourke, P. K. (1976). The effect of manifest construction activities on survival of deciduous trees in New Castle County, Delaware. M.S. Thesis, University of Delaware, Newark.

20

Chemical Injury

INTRODUCTION

A large number of chemicals used by people are toxic to plants. The most common groups of chemicals that are toxic to trees will be discussed under the following categories: deicing compounds, chemicals that cause spray injury, herbicides, and gases from underground pipes. Toxic gases, smoke, and aerosols found in the atmosphere are considered to be air pollutants and will be discussed in Chapter 22.

DEICING CHEMICALS

Sodium chloride (NaCl) and calcium chloride ($CaCl_2$) are the two most commonly used chemicals to melt ice and snow on sidewalks, driveways, and highways. Both these chemicals are sometimes applied together. Sodium chloride, however, is used most commonly, either alone or in combination with abrasives such as sand, cinders, washed stone, or slag screenings. Calcium chloride is used most commonly in extreme cold, below 20°F (–7°C), because it releases heat when it contacts water and can melt snow and ice at much lower temperatures than can sodium chloride. Calcium chloride, however, is more expensive and more difficult to handle than sodium chloride.

Deicing salts have been applied to highways in many areas of the United States at very high rates. Smith (1975) reports amounts as high as 2000 lbs (900 kg)/mile/storm on a two-lane highway in the northeast. Salt in runoff enters the soil around the roots of roadside trees and shrubs (Fig. 20.1). Salt spray from traffic is blown by winds or driven by turbulence to roadside plants where it coats the foliage of evergreens and also the stems and branches of all woody plants.

The exact effects of deicing salts in the soil on roots are complex but the salts are known to make water and essential nutrients difficult to absorb by the roots.

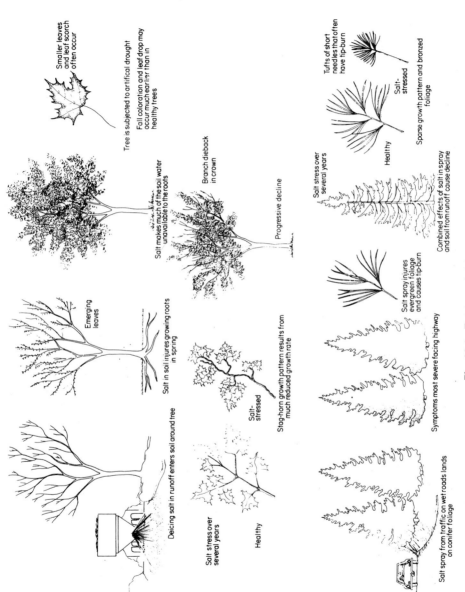

Fig. 20.1 Deicing salt injury.

Smaller leaves and leaf scorch often occur

Tree is subjected to artifical drought

Fall coloration and leaf drop may occur much earlier than in healthy trees

Salt makes much of the soil water unavailable to the roots

Branch dieback in crown

Progressive decline

Emerging leaves

Salt in soil injures growing roots in spring

Salt-stressed

Stag-horn growth pattern results from much reduced growth rate

Deicing salt in runoff enters soil around tree

Salt stress over several years

Healthy

Tufts of short needles that often have tip-burn

Salt-stressed

Healthy

Sparse growth pattern and bronzed foliage

Salt stress over several years

Combined effects of salt in spray and soil from runoff cause decline

Salt spray injures evergreen foliage and causes tip-burn

Symptoms most severe facing highway

Salt spray from traffic on wet roads lands on conifer foliage

Water is tightly held by the salt ions and more energy is required for the roots to absorb water. When sufficient water cannot be absorbed by the roots to meet the needs of the plant a water deficit occurs. This condition of salt-induced water deficit is sometimes termed a "physiological drought" because there is water in the soil but it is unavailable to the tree. The plant may respond to physiological drought by absorbing salts in an attempt to balance the soil concentration internally, thereby allowing absorption of water to occur. This response is thought to be an important mechanism for salt tolerance by some plants. This adjustment in metabolism usually requires considerable expenditure of energy, which must be drawn from other activities of the tree. Some plants use so much energy adjusting to soil salinity they stop growing, decline, and eventually die. Salt-tolerant plants appear to be able to adjust to increased soil salinity with little or no decrease in growth. However, the increased concentration of certain salts is also toxic to many plants. Both the sodium and chloride ions have been shown to be toxic. In these cases ability to absorb salts does not enable the plant to avoid salt injury.

Nutrient balance in the plant can also be affected by salt in the soil. The high concentration of sodium in salt-contaminated soil makes potassium less available to the roots. Potassium and sodium have similar chemical properties but only potassium is useful to the plant. High concentrations of sodium in soil can, therefore, result in preferential absorption of sodium instead of potassium. A similar deficiency can sometimes occur with phosphorus when plants are grown in soils with high concentrations of chlorides.

Salt spray is injurious to foliage, twigs, and buds of trees and shrubs. Chemicals within the spray are thought to enter the plant tissues, where they reach toxic levels. The foliage of broad- and narrow-leaved evergreens is the part most commonly injured by salt spray. However, injury can also occur to buds and on twigs and branches with thin bark. Trees are usually injured much more severely on the side of the tree facing a highway than on the side facing away from it. Affected trees often develop a one-sided appearance with most growth on the side away from the highway. In general, injury from salt spray is most severe on trees nearest the highway and decreases rapidly away from it.

The symptoms caused by salt injury are varied due to the different ways salt reaches plants and the different responses of plants to salt stress. Energy starvation from absorbing water with high salts will result in a general decline. Early fall coloration, small leaves, heavy seed load, and progressive twig and branch dieback are some of the common symptoms associated with this type of salt injury. Inability to absorb enough water to meet the needs of the crown will result in drought symptoms such as interveinal necrosis, small leaves, early leaf drop, and twig and branch dieback.

Salt toxicity often results in tip or marginal necrosis, which may eventually progress to a decline. In many cases, especially on salt-sensitive species, several types of injury may occur at the same time. Salt injury may also predispose the tree to other stress factors, such as drought and cold-temperature injury.

Accurate diagnosis of salt injury is best achieved with chemical analyses of both soil and affected plant tissue (see the section on soil testing in Chapter 17, Soil Stress). Both analyses are needed because some plants don't absorb salts. Knowledge of the amounts of deicing salt used on highways in an area may also be of considerable help. Salt in runoff and spray may not be the only sources of deicing salt. Storage areas for deicing salts or sand mixed with salts can also be sources of salt runoff. Waste from snow removal can also be an important source of salts.

There are also other sources of salt that can occasionally injure plants in addition to deicing compounds. In areas close to the shoreline injury from salt spray is occasionally a problem. It is most severe when high winds blow inshore during storms. Injury can often be seen several miles inland and is most severe on the seaward sides of the trees. Excessive use of fertilizers around trees and lawns can also result in increased soil salinity and consequent injury. Another source of salt injury that was reported by Feder (1976) is spray drift from cooling fountains of power plants that use seawater. Evaporation from the cooling fountains causes an increase in salinity of the recycled seawater. Under proper atmospheric conditions a salt-laden cloud is formed and drifts to adjacent plants coating them with salt. This type of injury has since been found around several salt-water-fed cooling systems on both fossil fueled and nuclear power plants.

Injury from salt can be prevented most effectively by keeping salt runoff and salt spray away from shrubs and trees whenever possible. Homeowners should avoid the use of salt around homes and rely instead upon abrasives for deicing on walks and driveways during the winter. In most cases, however, salt application is by local and state highway crews and is not under the direct control of the homeowner. Injury can be minimized by preventing as much runoff and spray as possible from contacting shrubs and trees. Construction of ditches to carry runoff away from trees will decrease injury from soil contamination. Salt spray can be blocked, at least partially, by placing barriers or screens around the trees during the winter months. Plywood, burlap, tar paper, and plastic have been used successfully to build temporary screens. Application of antidesiccants on foliage of evergreens also has been shown to decrease injury from salt spray.

Salt injury can also be avoided by proper planning of the trees and shrubs that are to be planted. In areas where highways are heavily salted and/or carry high speed traffic, plantings should be placed at least 30 ft (9 m) from the highway. Preference should be given to the highest areas, in relation to the road, to avoid runoff and as much spray as possible. Salt tolerance should be considered before any tree or shrub is planted. A brief list of the salt tolerance of some common trees and shrubs is given in Table 20.1. A more extensive list is presented by Dirr (1976). It is best to select only salt-tolerant species in areas where salt is likely to contact plantings (roadside, most urban areas, and in suburban and rural areas near major highways).

Salts cannot be prevented from entering the soil and/or contacting the foliage in many cases. Therapeutic treatments can be applied in these cases to remove

TABLE 20.1

Salt Tolerance of Some Common Trees and Shrubs

Tolerant	Sensitive
Shrubs	
Adam's needle	Arctic blue willow
Autumn elaeagnus	Boxwood
Bayberry	Japanese barberry
Beach plum	Multiflora rose
Buffaloberry	Van houtle spirea
California privet	Viburnums
Matrimony vine	Winged spindle tree
Pfitzer juniper	
Rugosa rose	
Tartarian honeysuckle	
Evergreen trees	
Austrian pine	Balsam fir
Colorado blue spruce	Canadian hemlock
Japanese black pine	Douglas fir
Pitch pine	Eastern white pine
Red cedar	Red pine
White spruce	
Yews	
Deciduous trees	
Big tooth aspen	American elm
Black cherry	American linden
Black locust	Boxwood
Box elder	Ironwood
Burr oak	Little-leaf linden
English oak	Red maple
Golden willow	Shagbark hickory
Green ash	Silver maple
Honey locust	Speckled alder
Quaking aspen	Sugar maple
Red oak	
Russian olive	
Siberian crabapple	
Siberian elm	
Weeping willow	
White oak	
White poplar	

the salt from the soil or plant tissue. Affected plants can be treated by leaching the salts from the soil with fresh water. This can be done at any time when the ground is not frozen but is most effective soon after a period of salt contamination. Removal of salt during early spring will allow a better flush of root and shoot growth. Leaching is accomplished the same way as watering (see Chapter 16, Moisture Stress). Several water applications may be necessary to remove most of the salt depending on the original levels. Salt on foliage from spray can also be removed in a similar manner by washing the foliage with fresh water. Since evergreen foliage is usually cleaned of deicing salt by spring rains, this treatment is usually applied only to foliage of trees near the sea coast following storms during the summer. Fresh water sprays must be applied soon after the storm if injury is to be prevented. Activated charcoal has been recommended to neutralize salts in the soil. Charcoal should be carefully worked into the top 6 inches (15 cm) of the soil under the dripline as soon as possible after salt contamination has occurred. Gypsum (calcium sulfate) has also been recommended to help remove salt from heavier soils where salt may be bound to the soil and resistant to leaching. Calcium ions in the gypsum are thought to replace sodium in soil colloids, thereby allowing more effective removal of sodium by leaching. Gypsum is applied to the soil surface at the recommended rate of 100 lb/1000 ft^2 (45.4 kg/10^2 m).

HERBICIDE INJURY

Herbicides or "weed killers" are chemicals that can cause the death of both woody and herbaceous plants. These chemicals are quite useful in controlling undesirable plants around agricultural crops, in recreation areas, and around homes. When used properly, herbicides are effective and only kill the target plants. In recent years, however, widespread application of herbicides has often resulted in their misuse with subsequent injury to nontarget plants (Fig. 20.2).

Injury from herbicides on individual trees and shrubs is influenced by a number of factors, such as the material used, the rate of application, the time of application, the species of plant, and the temperature. Injury can occur from direct contact of materials on all aboveground parts of the plant or through root absorption and translocation. Some herbicides cause very little or no injury to trees and shrubs while others (brush killers) are designed specifically to kill woody plants. The type of herbicide used, therefore, usually determines the probability of herbicide injury. Most herbicides can be classified as one of the following: preemergence, postemergence, contact, and soil sterilant.

Preemergence Herbicides

These materials are used most commonly on lawns to prevent the growth of weeds, particularly crabgrass. They are usually available in granular form, either alone or in combination with fertilizers, and are normally applied with a lawn spreader. As a group they are low in toxicity and most can be used safely on turf around trees and shrubs. Injury, however, has been found with high rates of simazine [2-chloro-4,6-bis(ethylamino-s-triazine)] and dichlobenil (2,6-dichlorobenzonitrile). The most common symptom of injury from these herbicides is a narrow chlorotic band around the entire leaf margin.

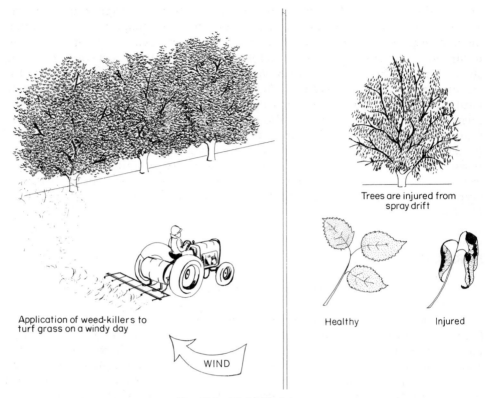

Trees are injured from
spray drift

Application of weed-killers to
turf grass on a windy day

WIND

Healthy   Injured

Fig. 20.2   Herbicide injury.

Postemergence Herbicides

These herbicides are applied to undesirable plants that are already established. They fall into two general groups: those that act like plant growth regulators or hormones and those that do not.

Hormone-Type Herbicides   Hormone-type herbicides are chemicals that resemble closely the structure, and therefore the mode of action, of certain plant hormones. The most common of these materials are 2,4-D (2,4-dichlorophenoxyacetic acid) and 2,4,5-T (2,4,5-trichlorophenoxyacetic acid), both of which closely resemble auxin (indoleacetic acid). Some other commonly used hormone-type herbicides are silvex [2-(2,4,5-trichlorophenoxy) propionic acid], MCP [2-(4-chloro-2-methylphenoxy) propionic acid], and dicamba (3,6-dichloro-o-anisic acid). Although there are a large number of hormone-type herbicides their mode of action and consequently their effect on plants are similar. These materials all cause abnormal growth of plant parts. Some common symptoms are curled leaves, twisted petioles, and distorted shoots and twigs (Fig. 20.3). Some of

Fig. 20.3 (A) 2,4-D injury on white ash. (B) Close-up. Note curled leaves typical of injury from hormone-type herbicides.

these herbicides, like 2,4-D, break down soon after application while others, like dicamba, may persist in soil for long periods. Persistent materials are most likely to move downward into soil and injure roots of woody plants.

Nonhormone Herbicides These materials do not resemble the chemical structure of plant hormones but disrupt essential steps in metabolism and thereby cause injury or death. One example of this group is aminotriazole (3-amino-1,2,4-triazole), which blocks chlorophyll synthesis and causes foliage to turn white to pinkish-white. The affected plant cannot utilize the energy from sunlight with chlorophyll and subsequently dies. Another common herbicide of this type is dalapon (2,2-dichloropropionic acid). Both aminotriazole and dalapon are translocated from roots to the leaves and are termed systemic herbicides. Injury can occur, therefore, in the upper branches of a tree following application of a systemic herbicide to an undesirable plant growing near the periphery of the tree's root zone.

Contact Herbicides

Contact herbicides are similar to the nonhormone herbicides in their mode of action but produce rapid killing of plant tissues on contact with little translocation. These herbicides are usually not applied near trees and shrubs but may cause injury from spray drift. Injury will usually appear as small necrotic spots where drift contacts foliage. Some herbicides in this group are sodium arsenite, paraquat (1,1-dimethyl-4,4'-bipyridinium ion), and cacodylic acid (dimethylarsinic acid).

Soil Sterilants

Chemicals applied to soil that kill all living organisms are known as soil sterilants. They may be volatile liquids or gases that do not persist in the soil or they may be long-lasting chemicals. Temporary soil sterilants are most commonly used prior to planting in nurseries and in various other agricultural operations. Sterilants are sometimes used to treat soil suspected of containing infectious root disease organisms. Soil sterilants are also used to prevent pathogen spread via root grafts. Temporary soil sterilants escape from the soil as gases and the treated area can usually be planted in a few days after application. These materials can kill roots while in the soil and caution must be exercised during application near existing trees. Some common materials in this category are vapam (sodium N-methyldithiocarbamate), methyl bromide, and vorlex [mixture of methyl isothiocyanate (20%) and chlorinated C_3 hydrocarbons (80%)]. Semipermanent soil sterilants can continue to cause considerable root injury to woody plants long after they are applied and can be carried by surface water to cause injury to surrounding plants. Some common materials of this type are atrazine (2-chloro-4-ethylamino-6-isopropylamino-s-triazine), borate–chlorate mixtures, bromacil (5-bromo-3-sec-butyl-6-methyluracil), diuron [3-(3,4-dichlorophenyl)-1,1-dimethylurea], karbutilate [m-(3,3-dimethylureido) phenyl-tert-butyl carbomate), monuron [3-(p-chlorophenyl)-1,1-dimethylurea], prometryn [2,4-bis(isopropylamino)-6-methylthio-s-triazine], and simazine. These soil sterilants are quite toxic to most woody plants, and their use near shade trees requires extreme caution.

Diagnosis of Herbicide Injury

Injury from herbicides is difficult to diagnose accurately and usually requires more than symptom recognition on injured plants. Many noninfectious and some infectious diseases can cause symptoms that appear similar to herbicide injury. There are, however, typical patterns of injury that are most commonly associated with herbicides, such as twisted and distorted leaves. Injury of a similar type to several species of trees, and sometimes also on ground vegetation in the same area, is usually a good indication of herbicide damage, but such injury is not restricted to herbicides. History of herbicide use in a location is also essential information in helping diagnose injury. However, this information may not be readily available and may even require a little detective work by the investigator. Herbicides are sometimes combined with other materials, such as fertilizers to obtain "weed and feed" mixtures, and the applicator may not be aware of the exact nature of the materials that were used. In these cases, ask the homeowner to examine the labeled contents of the container. Also ask the homeowner the amounts and mixtures used, the weather conditions during application, and the exact areas treated. Determine if there are any areas near the site of injury that are routinely treated with herbicides, such as areas around power lines, railroad and highway rights-of-way, and certain commercial installations. Any areas of open ground that are conspicuously "weed free" have most likely been treated with herbicides. This type of information will also be invaluable in attempting to diagnose other forms of chemical injury.

Control of Herbicide Injury

Most herbicide injury can be prevented merely by following directions on the label. Avoid excessive application rates and exercise care in application to avoid nontarget plants with spray and spray drift. Drift can be minimized by spraying only during periods of relative calm, by using coarse sprays or foams to avoid fine droplets, and by keeping spray nozzles as close to the target plant as possible. Equipment used for herbicide application should not be used for other purposes such as fungicide, insecticide, or fertilizer applications. If this is not possible, application equipment must be thoroughly cleaned before other materials are used. Avoid excessive applications of persistant herbicides that may build up in the soil. Reapply such materials only when weeds reappear.

When injury has occurred from herbicides it may be found in a range from mild injury requiring no treatment to the rapid killing of the plant in severe cases. Before any therapeutic treatment can begin it must be known whether the injury came from a foliar spray or from soil applications, and whether the herbicide is persistent or nonpersistent. If the herbicide is nonpersistent it can be assumed that no further injury will occur and treatments can be aimed solely at helping the tree to recover. If, however, the herbicide is persistent then efforts must also be made to detoxify the soil to prevent further injury. Most herbicides can be detoxified with activated charcoal applied over the tree roots and worked into the soil. The amount of charcoal needed to detoxify the herbicide is related to the amount of active ingredients applied. Smith and Fretz (1975) give detailed instructions on the calculation of the amount of charcoal needed to detoxify herbicides. Minor foliar injury generally does not require any treatment but defoliation or severe foliar injury may cause considerable stress. Application of balanced fertilizer in spring or fall will help the tree replace nutrients lost through foliar injury. Root injury from herbicides should be treated like other forms of root injury with branch thinning and phosphate fertilizer application to encourage root growth. Although herbicide injury may often appear severe, especially on foliage, many trees recover and resume normal appearance during the next growing season.

Fig. 20.4 Spray injury from 12% DDT on (A) gray birch; (B) American elm; (C) red maple; (D) white ash. (Photo courtesy of William B. Becker, University of Massachusetts, Amherst.)

INJURY FROM OTHER PESTICIDES

In addition to herbicides, other pesticides can also cause injury to plants when used improperly. Antibiotics, fungicides, insecticides, miticides, and nematicides are valuable chemicals that are sometimes needed to keep plants healthy but must be applied properly to be effective against pests and nontoxic to plants. The pests that these materials are designed to control are biologically similar in many respects to the plants they are designed to protect. Therefore, there is often a fine line between controlling the pest and causing injury to the treated plant. Most pesticide application does not result in any plant injury, but there are certain conditions that are often associated with pesticide injury (Fig. 20.4). Some of the most common of these conditions are incompatibility of mixtures, intolerance of materials, unfavorable environmental conditions, and poor health.

Incompatibility of Mixtures

Several types of pesticides are commonly mixed together in a spray tank to achieve broad spectrum control against several types of pests with a single application. When properly used this practice enables fewer sprays and more effective and economic use of pesticides. Not all pesticides, however, can be mixed with each other. Sometimes these materials react with each other and their chemical structures are changed. The products of these mixtures are usually much less effective than either material alone and often may be injurious to plants (phytotoxic). Such negative interaction of pesticides is termed incompatibility. There are charts available commercially and at most state universities that list the common compatible and incompatible mixtures of most pesticides. These lists, no matter how extensive, do not include all the available formulations of the numerous pesticides that are available today. Although pesticide manufacturers extensively test mixtures with their pesticides, it is not too difficult to prepare mixtures whose efficacy or possible phytotoxicity are unknown. When preparing mixtures, adhere carefully to recommendations on the label and only combine materials considered safe on an incompatibility chart.

Intolerance of Materials

Most pesticides can be applied to a large number of plant species without causing any harm. There are, however, some species of plants that are sensitive to certain materials. If pesticide intolerance of any species is known, it is usually written on the label and should be checked before application. Injury may often occur from dripping of spray from trees onto sensitive plants or from spray drift. There are also some materials that must be applied precisely according to labeled instructions or they will often cause injury. One of the most common groups of pesticides in this category is the dormant oils. Petroleum products, such as oils, are usually phytotoxic and if plant injury is to be avoided dormant oils should not be applied near or after bud break. Many thinbarked trees, such as dogwood, Lombardy poplar, magnolia, and redbud, and many evergreen species, such as junipers, are often injured from dormant oils. Repeat applications are usually injurious to most species of trees and shrubs.

Unfavorable Environmental Conditions

Environmental conditions can alter the effectiveness of pesticides and in some cases can cause plant injury. Pesticides are effective over a wide range of temperatures but they sometimes decompose when applied or stored at extremely high or low temperature. When pesticides are applied at temperatures above 90°F (33°C) foliar injury often occurs. At temperatures below freezing some materials, especially those containing emulsified oils, can separate and the breakdown products can cause plant injury. Some pesticides kept in solution on leaves during periods of high humidity may eventually decompose and cause plant injury. Most environmentally induced pesticide injury can be avoided by checking weather forecasts in advance of application and by spraying only when temperature extremes are not expected and when applied materials will dry in a few hours.

Poor Plant Health

Although plants in poor health or vigor are often prime candidates for vigorous pest control programs, these stressed plants may be additionally weakened by pesticides. When these chemicals are applied they often induce additional stress. Pesticides in general are slightly toxic to plants but a healthy plant can easily overcome the stress and is greatly benefited by the protection pesticides afford. Plants in poor health, however, may not be able to tolerate this additional stress and display symptoms of injury such as defoliation. When trees or shrubs are obviously stressed the major therapeutic efforts should be directed to overcoming the stress and returning the plants to normal health. If a minor problem occurs, such as mild insect infestation or fungus infection of the leaves, pesticides may not be required. If there is, however, need for protection by pesticides, apply only the minimal dosage and number of applications to control the problem.

UNDERGROUND GAS INJURY

Leaks from underground gas pipes can result in severe injury and death of both woody and nonwoody plants. Two types of gas have been used by utilities: manufactured gas and natural gas. Manufactured gas is generated by the incomplete combustion of coal or coke and consists mainly of hydrogen, carbon monoxide, and methane. Manufactured gas, however, is highly toxic to both plants and animals and its use has been largely discontinued. Natural gas is a by-product of the formation of petroleum, usually occurring in the ground with or near oil deposits. Natural gas is piped from the ground directly to the consumer and is chiefly a mixture of methane, ethane, and nitrogen. Since the use of manufactured gas is now quite rare, the underground gas found in pipes under streets is primarily natural gas. Although natural gas is not toxic to plants, it has been shown to cause changes in the soil environment around the roots that are detrimental to many woody and nonwoody species.

Natural gas in underground pipes is under pressure. This pressure may be low (as in pipes entering residences) or high (as in pipes along the main transmission line). The gas has a drying effect on caulking between pipe sections that may result in cracks with subsequent gas leaks at the joints. Major leaks are usually detected quite readily by a characteristic odor. Minor leaks, however, often go undiagnosed for several years. The amount of gas that will be lost through a leak is directly proportional to the pressure of the gas.

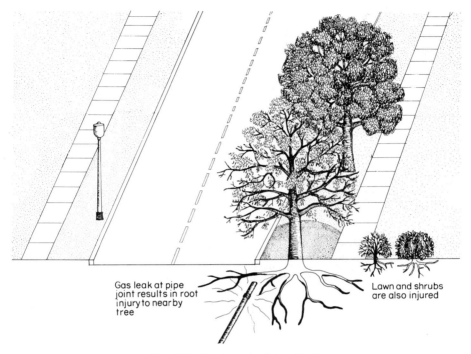

Gas leak at pipe
joint results in root
injury to nearby
tree

Lawn and shrubs
are also injured

Fig. 20.5 Underground gas injury.

When natural gas enters the soil from a gas leak it is usually at a level below which most plant roots grow [2 to 6 feet (0.6 to 1.8 m)]. The gas subsequently moves upward through the soil to a zone in which roots are growing (Fig. 20.5). The gas immediately begins to displace the oxygen in the soil and is also thought to cause soil drying. Natural gas in the soil also causes changes in soil microorganisms and sometimes in soil nutrients. Large populations of methane-consuming bacteria utilize any available oxygen and make conditions anaerobic. Bacteria that can produce hydrogen sulfide usually follow and cause the "sour gas" smell often associated with natural gas leaks. Since roots need abundant oxygen to respire and the metabolic activities of anaerobic bacteria are toxic, tissue in the affected roots soon begins to die after gas enters the soil. In addition to these microbial changes, the concentration of certain elements such as iron and manganese has risen, in some cases, to toxic levels. The magnitude of injury will depend upon the amount of gas in the soil and the length of time the leak has continued (Fig. 20.6).

Injury from natural gas leaks is often misdiagnosed because the symptoms can be confused with those from a large number of other disease problems. There are, however, patterns of injury that are most commonly associated with natural gas leaks in the soil. Injury to trees often occurs in a circle radiating from the

Fig. 20.6 (A) Sugar maple (center) dying from underground natural gas leak. Sugar maple at right is also beginning to exhibit symptoms from the leak. (B) Close-up of dying sugar maple. Note extensive bark killing. (Photos courtesy of Shade Tree Laboratories, University of Massachusetts, Amherst.)

source of the leak at the center. Most severe injury will occur nearest the leak and decrease progressively away from it. Herbaceous ground vegetation is often injured or killed as well as trees and shrubs. Severe gas leaks may cause rapid killing of all roadside trees on a section of a street while small gas leaks may be characterized by a slow decline of only one or two trees. Soil excavated around the roots of trees suspected to be injured by gas will often have a "sour gas" smell and will be very dark or black. It is, however, most important to be certain that underground gas lines are present before proceeding with diagnosis of gas injury.

Accurate diagnosis of underground gas injury can be best achieved by detecting gas in the soil with a gas detection meter such as the Vapotester produced by Scott Davis Instruments in Lancaster, N.Y. To use this instrument holes are made with a crowbar at least 20 inches (50 cm) deep in the soil around plants suspected to be injured by gas. A hollow pipe, connected to the gas detection meter, is inserted into the hole and air is drawn from the hole into the meter via an aspirator bulb. If the sample air contains a combustible gas such as natural gas, the meter will give a positive reading and also indicate the amount of gas present. Often, several holes are needed before the leak will be successfully detected. Most utility companies are eager to stop leaks and will check the soil around roadside trees where a leak is suspected, especially when the request is made by an arborist or municipal tree official. Many tree officials and some arborists have these meters to enable more rapid diagnosis of gas leaks.

Injury from underground gas leaks can sometimes be prevented by testing soil before planting trees or shrubs in the vicinity of gas transmission pipes. When gas injury has already produced easily detectable symptoms it is often fatal to plants, because the roots are usually dead. Sometimes around minor leaks only a portion of the root system has been damaged and a tree may be saved by prompt treatment. Repair of the leak after positive gas detection in the soil is the first priority for both human safety as well as for the health of the tree. If a gas leak is detected the root system around the buttress area and at various points under the drip line should be examined to determine the amount of viable roots that remain. If more than 50% of the root system is alive, therapeutic treatments may increase chances of survival. Soil under the drip line of affected trees should be aerated to a depth of at least 12 inches (30 cm). This can be accomplished most effectively by forcing compressed air into the soil through a soil needle. Aeration treatment will flush any remaining gas from the soil and accelerate the return of soil to its normal aerobic condition. The crown should also be thinned at least 10% and all dead and dying branches removed. Balanced fertilizer should be applied to encourage root growth. Soil aeration should also be performed before replanting around trees or other ground vegetation that were killed from underground gas leaks.

Another source of underground gas is the gas contained in reclaimed landfills. The soil in such areas often contains high levels of methane gas as a result of decomposing refuse. When landfills are converted into parks or recreation areas, failure of trees and shrubs can sometimes be attributed to underground gas injury. Detection of gas in landfill soil is similar to that described earlier. Treatment of affected trees may not be feasible since the source may not be stopped. Planting of trees and shrubs may have to be avoided until soil gas decreases below toxic levels.

MISCELLANEOUS CHEMICAL SPILLS

Accidental spills of chemicals that are toxic to plants sometimes results in injury to shade trees. Chemical spills usually injure roots, but occasionally the aboveground portions of the tree are injured by direct contact. Spills of volatile chemicals and gases will be discussed in Chapter 22, Air Pollution. There are, however, a large number of such toxic chemicals commonly used by humans. The identification of the offending chemical is often quite difficult, and sometimes the injury is confused with numerous root disease problems. Certain chemicals, such as automobile antifreeze, a variety of petroleum products, and water from chlorinated pools, appear to be most frequently associated with injury to woody plants. These chemicals will serve only as examples of the general types of chemical spills that occur.

Antifreeze often "boils over" and spills from radiator exhaust hoses of parked cars on hot summer days. Antifreeze, which often has an ethylene glycol base, is quite toxic to plants and such spills can result in severe injury or death, especially to small trees and shrubs. Injury from antifreeze can often be diagnosed when injury is limited to the plants that surround parking areas and when other factors, such as salt, have been eliminated. This type of chemical spill may be prevented by erecting curbs or fences that prevent the front end of cars from extending over the soil around plants.

Petroleum products, such as gasoline, kerosene, heating oils, lubricating oils, and most greases, are toxic to plants. If these materials are spilled on or around woody plants, severe injury or death is likely to occur. Spills of petroleum products are most common where their use is widespread, such as around service stations, but similar spills may occasionally occur around the home. Another common source of chemical injury to plants around the home can result from spillage of home heating oil during delivery. Injury only around the oil intake spout is usually a clue to an oil spill. Disposal of petroleum products should be done only at designated landfill areas where these products will not cause plant injury and also will not pollute surface water. Injury from petroleum products can usually be diagnosed from examination of soil around the affected plant. Contaminated soil usually retains some odor of the petroleum product. If soil is placed in a covered bottle for a few hours or overnight at room temperature the petroleum odor may be easily detected. Since petroleum products do not mix with water, another diagnostic procedure is to add some soil to water, shake, and allow the mixture to settle. If oil droplets from on the top of the water, this indicates that some petroleum product was in the soil.

Chlorine is a toxic gas that sometimes causes injury to plants as an air pollutant. Chlorine, however, is also added to the water in swimming pools to kill bacteria and algae. Sometimes during high winds and storms pool water is splashed or overflows onto the soil around poolside trees and shrubs. Homeowners may also drain water from pools onto the same areas. Chlorinated water is toxic to most plants, but species vary in susceptibility. While only very sensitive plants will be injured by one spill, most plants in runoff areas around the pool will be injured by repeated spills. If plant injury is confined to poolside vegetation and there is evidence of water runoff on the soil, chlorinated pool-water can be suspected as a cause of the injury. Prevention of this injury can be obtained by diverting overflows back into the pool and by draining any pool-water directly into sewers or into a dry well.

These examples of common chemical spills indicate the diverse nature of chemicals used by people that are toxic to plants. These examples also indicate a common association of chemical use and zones of injury. Although the number of such chemicals is large and continues to grow, the association of frequent use of a material and subsequent injury to plants nearby is the best key in diagnosis.

Whenever miscellaneous spills have already occurred the best therapeutic treatment, in most cases, is to flush the soil with fresh water. This treatment is most effective if done before symptoms appear. Water will dilute the toxic material and wash it from the soil. If injury is not extensive most woody plants can recover when the offending chemical is washed from the soil. In some cases chemicals that bind to the soil may require neutralizing with materials such as activated charcoal (see the section on herbicide injury). The exact nature of the chemical, its toxicity, and its solubility in soil should be known before applying any treatments other than water.

LITERATURE CITED

Dirr, M. A. (1976). Selection of trees for tolerance to salt injury. *J. Arbori.* **2,** 209–216.

Feder, W. A. (1976). Impact of saline mists on woody plants. *Proc. Am. Phytopathol. Soc.* **3,** 228 (abstr.).

Smith, E. M. (1975). Tree stress from salts and herbicides. *J. Arbori.* **1,** 201–205.

Smith, E. M., and T. A. Fretz. (1975). Chemical weed control in nursery and landscape plantings. *Ohio Co-op. Ext. Serv., Bull.* **MM-297.**

SUGGESTED REFERENCES

Daniels, R. (1974). Salt: ice-free walks and dead plants. *Arborist's News* **39,** 13–15.

Garner, J. H. B. (1973). The death of woody ornamentals associated with leaking natural gas. *Proc. Int. Soc. Arbori.* **49,** 13–16.

Hoeks, J. (1972). Changes in composition of soil near leaks in natural gas mains. *Soil Sci.* **113,** 46–54.

Holmes, F. W. (1961). Salt injury to trees. *Phytopathology* **51,** 712–718.

Holmes, F. W., and J. H. Baker. (1966). Salt injury to trees. II. Sodium and chloride in roadside sugar maples in Massachusetts *Phytopathology* **56,** 633–636.

Leone, I. A., F. B. Flower, J. J. Arthur, and E. F. Gilman. (1976). Landfill gases: A source of plant damage. *Prod. Am. Phytopathol. Soc.* **3,** 307 (abstr.).

Marsden, D. H. (1951). Gas injury to trees. *Trees Mag.* **12,** 20 and 24–55.

Meade, J. A. (1975). Street trees and herbicides. *J. Arbori.* **1,** 68–70.

Rich, A. E. (1971). Salt injury to roadside trees. *Proc. Int. Shade Tree Conf.* **47,** 77a–79a.

Shortle, W. C., and A. E. Rich. (1970). Relative sodium chloride tolerance of common roadside trees in southeastern New Hampshire. *Plant Dis. Rep.* **54,** 360–362.

21

Tree Maintenance

INTRODUCTION

Proper maintenance is essential to the health of shade trees because improper maintenance can kill trees. Tree maintenance is a demanding task that requires skill in arboriculture, botany, entomology, plant pathology, and many other related sciences. Care of trees should not be undertaken by untrained persons because the results will often be more detrimental to the tree than the original problem. Examples of improper maintenance have recently become more common, as homeowners and other nonprofessionals attempt to practice tree maintenance.

Most tree maintenance falls into one of the following general categories: planting, pruning, wound treatment, spraying for control of diseases and insects, cavity treatment, fertilizing, and watering. Each of these tree maintenance operations can be useful to the tree if performed properly or can cause harm to the tree if performed improperly. These maintenance operations will each be discussed in detail; the proper techniques will be briefly discussed, and the most common examples of unsound practice will be presented. The objective of this chapter is not to present an extensive guide to proper tree maintenance, which would require many chapters. The objective is to point out the major examples of injury to shade trees as a result of improper maintenance and show how such injury can be avoided.

TREE PLANTING

Techniques used during planting of a tree will determine whether a high quality shade tree will develop or whether a stunted or diseased tree will result. It is doubtful there can be a more critical operation in the life of a tree than transplanting. Critical steps in tree planting, however, may begin years earlier, especially with large trees, which must be prepared gradually. Tree health and preparation are critical to success in tree planting. Follow-up maintenance is equally critical. Proper planting of shade trees involves selection of healthy stock suitable for the planting site, preparation of site, correct placement of the tree in the soil, and follow-up maintenance.

Tree Selection

Many site conditions such as soil type, wetness, weather extremes, root space, and crown space should be considered before selecting a tree species. Large, fast growing trees should be avoided in confined locations such as near buildings, overhead wires, and other trees (Figs. 21.1 and 21.2). Planning ahead for the expected size of the shade tree at maturity avoids needless removal or disfigurement later (Fig. 21.3). The ideal growing conditions for most shade trees are well known. Many books and leaflets on all aspects of tree selection are available, and some are distributed at no charge or a nominal fee by various government organizations. Some of these are among the suggested references at the end of this chapter. However, the easiest and perhaps most effective way to determine the tree species that are ideally suited for a particular location is to check a nearby wooded area and determine which tree species occur naturally. Any trees with obvious stem cankers, numerous dead branches, or severe mechanical damage should not be accepted. It is, however, usually difficult or impossible to accurately determine the health of a tree before purchase. It is, therefore, critical to select plants from reputable nurseries that will guarantee the quality of their plant materials. Trees may also be stored 1 or 2 years in a "holding" nursery (Fig. 21.4) before outplanting to allow them to fully recover from transplanting and shipping in a location ideal for growth. Transplanting trees from woodlots may appear to be an inexpensive way to obtain a quality shade tree but often results in tree failure. Trees in the forest usually have a sparse and dispersed root system. Consequently, a considerable amount of root injury often occurs during transplanting and chances of success are severely diminished. Trees transplanted from the forest often need considerable follow-up maintenance to survive. Trees in a nursery are usually root-pruned and have a dense compact root system that can tolerate transplanting with less injury. Nursery trees are also able to resume normal growth and vigor shortly after transplanting with less chances of attack by secondary pathogens or insects. Forest trees, however, can be transplanted with increased success if the roots are cut at the periphery where the root ball would be formed. The roots are usually cut in stages, 50% from each quadrant each year, for two years prior to transplanting. This practice will decrease root injury during transplanting and induce a more compact root system.

Trees may be selected either as stock that is bare rooted or balled and burlapped. Evergreens, however, are moved most often with a soil ball around the roots due to intolerance of bare-root transplanting. Deciduous trees over 3 inches (7.5 cm) diameter at breast height should also be moved with a soil ball.

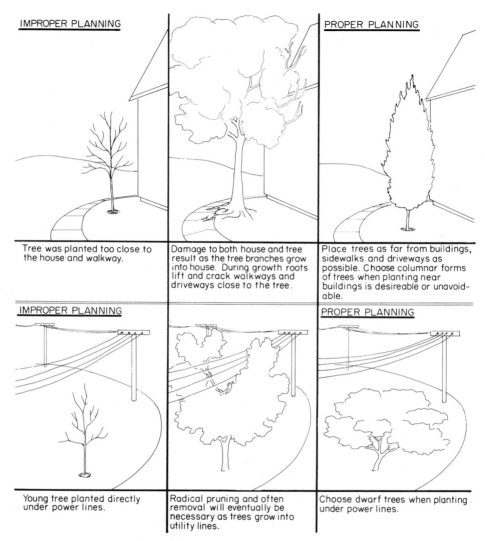

IMPROPER PLANNING

Tree was planted too close to the house and walkway.

Damage to both house and tree result as the tree branches grow into house. During growth roots lift and crack walkways and driveways close to the tree.

PROPER PLANNING

Place trees as far from buildings, sidewalks and driveways as possible. Choose columnar forms of trees when planting near buildings is desireable or unavoidable.

IMPROPER PLANNING

Young tree planted directly under power lines.

Radical pruning and often removal will eventually be necessary as trees grow into utility lines.

PROPER PLANNING

Choose dwarf trees when planting under power lines.

Fig. 21.1 Proper and improper planning.

Once trees have been dug, whether bare rooted or balled and burlapped, care must be taken to prevent drying of the roots or soil ball. If stock is received by mail from the nursery the trees should be checked for moisture as soon as they arrive. The planting stock must be kept moist and stored in a cool location under low light intensity. Planting should take place as soon as possible after the stock is dug or is received by mail. Trees must also be protected from drying throughout the planting procedure. Bare-root stock should be covered with wet burlap

Fig. 21.2 Proper planning of trees in a housing development. Norway maples are planted on the side without overhead wires and crabapples are planted on the side with the overhead wires but set back from the street.

and kept in shade until placed into the planting hole. In extreme hot spells, planting after sunset may be more successful. Only a few trees should be dug or removed from a storage area at one time, to decrease chances of drying-out of roots before planting. Only the number of trees that can be properly planted in a few hours should be dug or moved from storage.

Trees that are growing in containers can usually be handled more easily than bare-root or balled-and-burlapped stock and have become increasingly popular

Fig. 21.3 Improper planning. Ginkgos planted under wires required drastic pruning and disfigurement.

for planting. Transplanting from containers results in minimal root loss and
trees may be grown for extended periods before transplanting. Containerized
trees, however, are generally limited to trees under 10 ft (3 m) in height due to
the limited amount of available root space in the container; they eventually
outgrow the available root space and become "pot-bound." Containerized trees
are also susceptible to cold injury (see Chapter 15, Temperature Stress) and must
be protected from low soil temperature in winter.

Site Preparation

Survival of the transplanted tree depends on the ability of the root system to
resume normal root growth. Site preparation involves: (1) selecting the planting
site, (2) digging the planting hole, and (3) ensuring proper drainage.

Selecting the Planting Site   The location of a newly planted tree is usually
determined by the owner or is restricted by walks, underground utilities, other
trees, buildings, or other obstructions. A municipal tree official or arborist may
have to be concerned with underground water, sewer, and gas pipes as well as
underground electric power and telephone cables, each of which may enter from
the street at a different location. In addition, trees in a new subdivision must (if
possible) be located away from any future planned sidewalks or other future

Fig. 21.4 Holding nursery in West Springfield, Massachusetts where trees shipped from commercial nurseries are held for 1 or 2 years. Trees are outplanted to the roadside only when they are in prime condition resulting in a low mortality rate after planting.

construction projects that might require tree removal. The tree official or arborist must, therefore, not only select the best site for tree survival but must also determine where a planting hole can be placed. This may involve competition with other town officials, such as highway engineers, for the space. Information from several of these agencies is often required before a "safe" zone for planting trees may be selected.

Once a zone where planting can occur has been chosen, the best location for a tree within this area must be determined. Trees should be planted according to their tolerance of the soil conditions, which can often vary within the desired area of planting. Close inspection of a site for soil moisture and texture will often reveal the best specific location for a tree.

Digging the Planting Hole   There are many specific guidelines available for the preparation of a planting hole. In general the planting hole should be large enough to allow the entire root system to be placed into the hole and also allow room for a layer of soil that will be added below and outside the periphery of the roots. The soil layer will encourage root growth and should be made as wide as possible. A narrow planting hole will restrict root growth and may cause roots to turn and grow back around the trunk resulting in root girdling (see Chapter 17, Soil Stress).

Ensuring Proper Drainage The planting hole should be dug at least 1 ft (30 cm) below the root depth and then be partially filled with water. If the water drains from the hole in a few hours there is adequate drainage below the soil for the tree. If, however, the water remains pooled in the hole overnight there may be an impervious layer or hardpan below the planting hole. In most cases drainage can be improved by penetrating through this impervious layer by making a small hole 24 to 36 inches (60 to 90 cm deep) with a crowbar at the base of the planting hole. If improved drainage through the bottom of the planting hole cannot be achieved by this procedure, another planting site should be selected if available, or drainage away from the side of the hole, such as with drainage tiles, should be provided.

Placement of Transplanted Trees in Soil

Once the planting hole has been completed, a layer of light garden loam is added to the bottom. The loam is added to a depth that will place the planted tree a few inches (6–10 cm) above the soil line at which it was growing at the nursery.

Fertilizer may be added to the soil at this time. There seems to be no general agreement as to the best fertilizer to add at the time of planting. Some authorities on tree growth suggest that fertilizer applied at this time has no positive effect on tree growth and in some cases is detrimental and therefore, should not be added. Most researchers agree that the high concentrations of fertilizers that are added to established trees will be detrimental to a newly planted tree. Most also agree that root growth is critical to the success of a newly planted tree and low concentrations of balanced fertilizers are most beneficial to root growth.

The transplanted tree is then set on top of the loam (Fig. 21.5). If the tree is balled and burlapped, all cords tied around the ball should be removed at this time to prevent girdling as the tree grows. The wrap should be carefully rolled away from the root ball. A better idea of the original soil line on the root buttress area can usually be obtained at this time and adjustments for placement depth can be made. Wrap around the root ball that consists of a biodegradable material such as burlap may be left around the root system in the planting hole. Non-biodegradable plastics, such as plastic mesh burlap or plastic sheets, must be removed completely from the root ball. These materials may either girdle roots that have penetrated them, as in the case of plastic mesh burlap, or may suffo-cate roots and favor root pathogens, as in the case of plastic sheets. If in doubt on the biodegradable nature of the wrap, it is best to remove it completely. If the tree is bare-rooted the tree should be placed into the hole and the roots spread out away from the trunk. The roots should never be jammed or twisted into a small planting hole. If the size of the root system has been underestimated, the

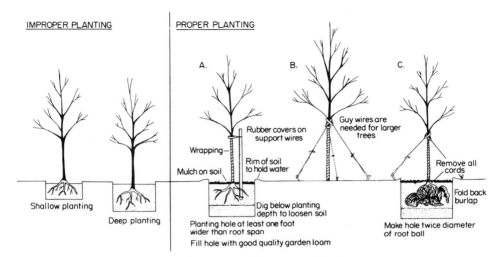

Fig. 21.5 Proper and improper planting.

planting hole must be enlarged. Roots that are twisted around the trunk or jammed against the side of the planting hole will often grow into the trunk and cause girdling.

The tree, whether balled and burlapped or bare-rooted, should be set in the ground at a depth 2 to 3 inches (6 to 10 cm) above the level it was growing at the nursery. In bare-rooted stock this level can usually be determined accurately by a soil-stained area on the bark just above the uppermost lateral roots. The tree is set above the original level of growth to allow for settling in the planting hole. Larger trees, greater than 4 inches (10 cm) diameter at breast height, usually require additional elevation during planting because of the increased weight of the soil ball. Failure to allow for settling of the root ball after planting has resulted in stress from ''apparent'' deep planting in subsequent years. Once the tree has been placed at the proper depth, soil is added around the roots to bring the level of the planting hole equal to that of the surrounding area. The soil in the planting hole should be padded down carefully but not compacted. A ridge of soil 1 to 2 inches (2.5 to 5 cm) high should be formed around the edge of the planting hole. This will retain rainwater or applied water around the roots.

There are many methods available for staking or guying a newly planted tree. All these methods, however, are primarily aimed at protecting a tree from wind throw until new root growth can secure the tree in the soil, and affording reasonable protection for the tree against vandalism and theft. Supports, therefore, must be secure enough to keep the tree in place but not so tight that injury occurs to the bark. Guy wires and ties from the tree to the supporting stake must be covered with some type of broad and elastic material, such as a section of garden hose, wherever they contact the tree. Bare wire, cord, or ropes that are sometimes tied around the tree by homeowners will soon dig into the bark, causing injury and eventually girdling. Frequent interaction with people or animals often necessitates more elaborate supportive and protective devices. In such areas small trees, less than 1 inch (2.5 cm) diameter, are often surrounded by cages of hardware cloth ~18 inches (45 cm) high in addition to staking. Wire bar cages approx 6 ft (1.8 m) high are sometimes placed around larger trees for protection against injury.

A layer of mulch 4 to 6 inches (10 to 15 cm) deep should be spread over the entire root system (Fig. 21.6). The mulch will enable the tree to retain moisture more effectively and will provide a good soil environment for root growth. A

Fig. 21.6 Proper planting of a 4-inch (10-cm) pin oak. Note trunk wrap, guy wires, with turn buckles for loosening and brightly colored flags to warn pedestrians, and mulch over the root area.

large number of attractive mulches are available, such as bark chips, crushed stone, gravel, pebbles, and wood chips. Mulching also decreases injury from lawn maintenance equipment and minimizes soil compaction. Mulches are also used to prevent weeds; however, people may be tempted to place plastic sheets under mulches to make them more effective in weed control around trees. These plastic sheets can suffocate the tree's roots and also favor the growth of root pathogens and therefore should not be used under mulches.

Deciduous trees greater than 1 inch (2.5 cm) diameter should be wrapped to protect against sunscald (see Chapter 15, Temperature Stress). Trees should be wrapped beginning from the soil line and continue to at least the first set of lateral branches. Kraft paper is often used for wrapping but plastic wrapping is now available. Wraps should be left in place for 2 years. Kraft paper usually deteriorates and falls off by this time, but plastic wrapping should be removed after the second year following planting. The string that ties the wrapping in place must also be removed before it strangles the enlarging trunk.

The soil around the tree should be thoroughly saturated by watering at the time of planting. This is most critical for trees planted during warm or windy weather where soil moisture loss will be rapid. Since balled-and-burlapped stock has already been kept moist or wetted prior to planting, watering is aimed primarily at preventing loss of water from roots into surrounding soil. Bare-rooted stock usually necessitates a large amount of soil added to the planting hole, and it should be wet to prevent drying out of the root system. Moisture is also needed in the soil to cause soil particles to fill any tiny gaps between rootlets and to come into intimate contact with small roots.

Trees should also be pruned during transplanting, to balance the loss of roots during digging. Excessive moisture loss, with subsequent drought stress, can often occur if transplanted trees produce abundant foliage but have lost numerous roots during transplanting. Trees are usually thinned 10–20% with most branches taken from the upper crown and secondary shoots. This will enable the tree to retain its natural form. Bare-rooted stock should be pruned more heavily than balled-and-burlapped stock. Trees transplanted from forests, open fields, or any natural locations should be heavily pruned (30 to 40%) due to the large amount of root injury that occurs during transplanting.

FOLLOW-UP MAINTENANCE

Care of trees for 2 years after planting is often essential to their survival. Newly planted trees must have continued proper growth conditions to become established. The most commonly performed follow-up maintenance is watering, fertilizing, and removal of guy wires or other supports from around the trunk. Watering and fertilizing will be discussed in following sections of this chapter.

Support wires cease to serve a useful function once the tree roots have become established in the surrounding soil. After 2 years support wires should be re-

Fig. 21.7 Trunk girdling of a hemlock by a guy wire (A) and girdling of a London plane from hemp cord used to secure tree wrap (B).

moved, but the tension of the supports should also be checked periodically after planting and loosened if the trunk growth results in excessive pressure around the support. If support wires are not removed (one of the commonest forms of neglect) the trunk growth will eventually girdle the tree as the pressure against the support kills the bark (Fig. 21.7). This problem is found most commonly in coniferous evergreens where the support wires are often hidden by foliage.

Pruning

Proper pruning can relieve the tree of dead and dying branches, remove disease inoculum, improve tree vigor, compensate for root loss, and help a valuable shade tree to retain its desired form. Pruning performed improperly, however, can cause excessive wounding, decrease the vigor of the tree, predispose it to insect or pathogen attack, and drastically alter its desirable form. Pruning is part of the normal maintenance program of the tree and should be done at least every 2 to 3 years. Pruning techniques are discussed in detail in Chapter 13, Wound Diseases.

Trees can be pruned at any time of year but spring pruning of some species, such as birches and maples, results in abundant sap flow or "bleeding" from the pruning cuts. Although this sap bleeding is not thought to be harmful to the tree the sweet sap is often colonized by bright colored fungi. The trunks of pruned trees may be covered with fungus growth below the wounds, which may be considered unattractive by the homeowner. The fungi dry up after leaf flush when bleeding stops and are washed from the trunks by spring rains. Wound dressings applied during sapflow frequently wash off and must be reapplied when bleeding stops.

Trees can be pruned for a variety of reasons. Pruning is most often done to remove all dead and dying branches, as well as branches growing into other trees or obstructions such as utility lines or buildings, or branches rubbing against each other. Pruning is also commonly performed to remove branches cankered by infectious pathogens from the tree (see Chapter 9, Stem Diseases). In addition, pruning is often performed to compensate for root loss that occurs during transplanting and during a variety of root stresses, such as construction injury or soil compaction. The form or growth pattern of the tree can also be modified by pruning. Pruning to achieve each of these purposes will, of course, involve removal of different branches but proper pruning techniques must be employed in all cases.

Pruning of live branches must not be excessive because the rapid loss of food-producing capacity may cause the tree to go into decline. Severe pruning can also cause sunscald, especially in evergreens. Thinning should be done gradually over a few years to allow the tree to adjust to branch loss. In the case of severe root injury or loss, for example, no more than 20% of the crown should be removed initially. The following years an additional 10% of the crown can be removed until a root : shoot balance has been restored.

Wound and Cavity Treatments

Treatment of pruning wounds and other types of mechanical injury is discussed in detail in Chapter 13. Cleaning and filling of cavities is also discussed in that chapter.

Spraying for Control of Diseases and Insects

Applications of antibiotics, fungicides, insecticides, miticides, and other pesticides for control of diseases and insects are often essential to maintain the health of trees and shrubs. Pesticides, however, when used improperly can cause considerable injury. The effects of improper application of pesticides are discussed in detail in Chapter 20, Chemical Injury.

Watering

Watering is sometimes needed for trees even after they have adapted completely to transplanting. Most shade trees are open grown and consequently have a high demand for water due to large losses of transpiration from the leaves in hot weather. Since many locations where shade trees grow restrict root growth, these trees are often susceptible to stress from moisture deficiency. Water should be applied only to the soil around the roots and not on the foliage and branches. Prolonged wetness on aboveground parts of the tree favors development of leaf and stem pathogens. Watering of established trees is discussed in detail in Chapter 16, Moisture Stress.

Fertilizing

Fertilizers can help trees replace soil minerals that are lost through leaching and absorbed by the roots. Decomposition of fallen leaves in the forest recycles most of the minerals taken up by the trees but the leaves around shade trees are usually removed. The soil around shade trees, therefore, can become deficient in common soil elements such as nitrogen, phosphorus, and potassium. Many soils around homes and along streets are also deficient in soil minerals and require amendment before vigorous growth of trees can be attained. Periodic fertilizing, therefore, is a common recommendation to maintain the health and vigor of shade trees.

Fertilizers are available in a wide variety of forms. Fertilizers may be organic such as manures, or may be inorganic mixtures of salts, such as ammonium phosphate or potassium nitrate. Organic fertilizers are beneficial, if they have been well rotted, because they enrich the organic content of the soil as well as provide essential nutrients. Green manures should be avoided because of the danger of root injury. Inorganic mixtures or "commercial" fertilizers are easy to handle and store, are readily available, and provide accurate information of their composition. Commercial fertilizers are rated as to their proportions of nitrogen, phosphorus, and potassium, respectively, by three numbers on the package (such as 5–10–10). Fertilizers containing each of these components are termed balanced. Organic fertilizer contains balanced minerals but the exact proportions and concentrations vary from lot to lot, and cannot easily be determined. Slow-release formulations of fertilizers, especially those designed for liquid soil injection, are widely used to fertilize trees and shrubs. This type of fertilizer is usually a suspension that contains urea–formaldehyde formulations of nutrients that are gradually released by the action of soil microorganisms for 6 to 12 months following application. Soil bacteria convert the most soluble nitrogen fraction— approximately one third of that applied—to an available form for the tree within 3 to 5 weeks after application; the remaining fraction is converted over the following 5 to 11 months. Since nutrients, especially nitrogen, are released gradually over approximately 1 year following application, slow-release fertilizers can be applied at any time of year when the soil is not frozen.

Slow-release fertilizers, as well as immediate-release fertilizers, can be applied through a grid of holes made under the drip line of the tree. The procedure of liquid soil injection, however, has been shown to be four times faster per tree than dry application of fertilizers in drilled or punched holes. Soil injection also allows the combination of fertilizer treatment and irrigation. The grid of holes offers an added benefit to the tree in the form of soil aeration.

Broadcast application of fertilizer on the ground has been shown to be an effective method of fertilizing trees where ground vegetation is absent or not of concern, as in a commercial tree nursery. Surface application of fertilizer on turf around trees, however, can cause "burning" injury to the turf or can result in excessive turf growth under trees compared to the rest of the lawn.

Foliar application of fertilizer has been used to supplement soil fertilization. Foliage fertilization has been most effective in helping to correct micronutrient imbalances, such as an iron deficiency, or to achieve a rapid but temporary "greening" of foliage. It must be emphasized that foliar fertilization cannot entirely substitute for soil fertilization, and its effects usually last only a few weeks. An advantage of foliar fertilization is that it allows fertilization during pesticide applications. Although most foliar fertilizers are compatible with most pesticides, it is recommended that the compatibility of all components be checked before application (see Chapter 20, Chemical Injury).

The rate of fertilization for trees and shrubs is subject to some variation in recommendations. Most rate recommendations are now made according to soil area under the approximate drip line. This type of calculation is considered more accurate and less likely to result in overfertilization than rates based on tree diameter. The danger of injury from overfertilizing is greatest in shrubs, small trees, and recently transplanted trees. Large trees are generally fertilized at a higher rate per diameter than shrubs and small trees. The widespread use of slow-release fertilizers, however, has decreased the chances of injury from over-fertilization, since all the nutrients are not immediately released into the soil.

Fertilizers should always be applied to trees that are in a well-watered condition. If the soil is very dry at the time of application, there is a danger that a high concentration of fertilizer may be rapidly taken up by the tree, causing a foliar "burn" injury. It is recommended that fertilization be performed only when the top 8 inches (20 cm) of soil is moist, or the trees should be watered to achieve this degree of soil moisture before fertilization is performed.

Tree Injection and Implantation   Direct application of antibiotics, fertilizers, fungicides, and insecticides into trees via injection or implantation has become a widely used method of controlling diseases and insects or of stimulating growth. Improvements in systemic chemical formulations used in injection and implantation have enabled arborists to control many disease and insect problems effectively without spray applications. In addition, some diseases, such as wilts and

Fig. 21.8 Beneficial effect of a copolymer acrylamide–acrylate superabsorbant gel on established live oak bare root transplants. Two and a half months after planting (A) without and (B) with gel. Five months after planting (C) without and (D) with gel. (Photos courtesv Aquatrols Corporation of America, Pennsauken, New Jersey.)

some cankers, that could not be controlled through spraying have been controlled by injection or implantation. Trees in urban settings with pavement over most of their root systems can only be effectively fertilized by trunk injection or implantation. Injection and implantation, however, are effective only when performed by trained professionals. Since systemic treatments require wounding, care should be taken when applying treatments to minimize any wound effects (see Chapter 13).

Wetting Agents Wetting agents have been used in the turf and landscape industries for many years to improve water management. Recently, wetting agents were shown to have a beneficial growth effect on trees and shrubs in Central Park, New York City (see Chapter 16, Moisture Stress). Wetting agents are chemicals that lower the surface tension of water and are considered to be surfactants. Soils contain areas of differing hydrophobicity (ability to avoid becoming wet). Rain water or irrigation water will seek the path of least resistance, wetting certain areas and leaving others as dry spots. Wetting agents break down areas of resistance to wetting in soils, allowing a uniform distribution and greater penetration of water within the soil. In addition to uniform wetting, the use of a wetting agent can result in improved drainage and aeration. Plant growth is dependent upon a uniform level of moisture in soils as well as adequate aeration.

Superabsorbent Gels Superabsorbent gels are polymers that have the ability to hold 50 to 600 times their weight in water. Depending upon the composition of the gel, 40 to 98% of this water is available to the plant and the soil. Chemically, superabsorbent gels are of two types: starch or synthetic. Starch-based materials are generally less structurally stable than synthetic materials. This means that starch polymers may collapse in the soil while some synthetic polymers are known to be still functionally effective after 2 to 3 years.

The utility of these products in landscaping lies in their ability to increase the water-holding capacity of the soil system and to provide a reserve of water for the plant after the soil has dried. Generally, superabsorbent gels are incorporated into the backfill and planting hole. These materials can be very beneficial under stress conditions, or in soils with low moisture-holding capacities, to aid in the successful establishment of plants (Fig. 21.8).

SUGGESTED REFERENCES

Anonymous. (1966). Trees for shade and beauty: Their selection and care. *U.S., Dep. Agric., Home Gard. Bull.* **117,** 1–8.

Anonymous. (1972). Transplanting ornamental trees and shrubs. *U.S., Dep. Agric., Home Gard. Bull.* **192,** 1–12.

Anonymous. (1975a). Pruning ornamental shrubs and vines. *U.S. Dep. Agric., Home Gard. Bull.* **165,** 1–16.

Anonymous. (1975b). Trees in your community. Electric Council of New England. W. Mass. Electric Co., Springfield, Mass.

Chater, C. S. (1961). Transplanting small trees and shrubs. *Univ. Mass., Ext. Publ.* **344,** 1–7.

Holmes, F. W., and A. W. Boicourt. (1976). Pruning trees and shrubs; root pruning. *Univ. Mass. Ext. Publ.* pp. 1–4.

Holmes, F. W., and H. E. Mosher. (1975). Fertilizing trees and shrubs. *Univ. Mass., Co-op Ext. Publ.* **382,** 1–3.

May, C. (1970). Pruning shade trees and repairing their injuries. *U.S. Dep. Agric., Home Gard. Bull.* **83,** 1–15.

Neely, D., and E. B. Himelick. (1966). Fertilizing and watering trees. *Ill. Nat. Hist. Surv., Circ.* **52,** 1–20.

Neely, D., E. B. Himelick, and W. R. Crowley. (1970). Fertilization of established trees: A report of field studies. *Ill. Nat. Hist. Surv., Bull.* **30,** 235–266.

Pirone, P. P. (1978). Tree maintenance, 5th ed. Oxford Univ. Press, London and New York.

22

Air Pollution

INTRODUCTION

Air pollutants that are toxic to trees can occur in several forms: invisible gases, particulates (smoke or dust), or aerosols (fine mists). Many people associate air pollution only with smoke but invisible gases are the most common form of pollution toxic to plants (phytotoxic). Air pollution is not restricted to urban and suburban areas but is also found in remote rural areas. Most air pollution results from electric power generation, cars and trucks, and a variety of industrial operations.

Most initial symptoms in plants from air pollutants appear on the leaves. Toxic gases enter the leaves through the stomates and react with the internal leaf tissues. The affected tissues are inhibited in their photosynthetic ability and may be killed. Aerosols and particulates land on the leaf surface and can react both chemically and physically with the leaf tissues.

Injury from air pollutants can be put into two major categories: (1) chronic, and (2) acute. Chronic injury occurs from the cumulative effects of long exposure (often several growing seasons) to toxic pollutants at low levels on sensitive species, or at moderate to high levels on tolerant species. Acute injury occurs after short exposure (a few hours to several days) to pollutants at high levels on tolerant species or at moderate to low levels on sensitive species. Acute injury occurs commonly following accidental leaks or spills of gas or volatile liquids. Chronic injury is difficult to diagnose because the foliar symptoms are usually mild or nonexistent and the trees decline over several years. Acute injury results in moderate to severe foliar injury and is the most frequently diagnosed type of air pollution injury.

Air pollutants may be classified according to their sources into two broad groups: (1) point source emissions, and (2) diffuse oxidants. Point source emissions come from stationary sources such as smokestacks or burning areas. Diffuse oxidants are formed over a wide area in the atmosphere from various chemical reactions with oxygen that are powered by sunlight. These diffuse pollutants are also termed photochemical oxidants.

The specific effects of air pollutants on plant tissue vary with the pollutant, host, time of year, and numerous meteorological factors such as temperature, relative humidity, wind, and solar radiation. It is, therefore, difficult to describe or photograph "typical" symptoms of air pollution injury on trees or any other plants. In addition, symptoms known to be produced on plants by air pollutants may also be produced by stress from moisture or temperature, nutrient disorders, and other stresses, as well as a variety of biotic factors such as bacteria, fungi, viruses, sucking insects, and mites. A thorough examination of the affected plant should be conducted to eliminate any of these other disorders before air pollution injury is considered. The use of photographs or descriptions of air pollution injury for diagnosis without knowledge of local meteorological conditions and nearby sources of emission can often be misleading. The source, the species sensitivity as correlated with symptom type, the nature, and the movement of air pollutants must be known before accurate diagnosis of injury can be achieved. In addition, sensitive chemical analyses are sometimes needed to detect air pollutants such as fluorides and chlorides that are toxic in low concentrations and may come from distant or dispersed sources.

Control of air pollution injury to vegetation is a difficult task. Sources may be far from the areas where injury occurs and commonly cross town, county, state, and sometimes national boundaries. Various federal, state, and local government regulatory agencies are involved in monitoring pollution levels in an attempt to reduce emissions. Strict enforcement of air pollution control laws, however, is difficult. Planting of species resistant to phytotoxic air pollutants, therefore, is the most widely used measure to minimize injury where pollutants are known to occur. Application of balanced fertilizers to increase vigor has been recently shown to decrease injury on trees after exposure to some common pollutants.

METEOROLOGY

Air pollutants, whether gases, particulates, or aerosols, are carried by air movements. Numerous meteorological factors, such as wind direction and velocity, temperature, relative humidity and precipitation, and geographic factors, such as mountains, valleys, oceans, and lakes, govern the movement of air. These factors are critical to the understanding of air pollution injury on plants since they modify the concentration, and in some cases the chemical structure, of the air pollutants. The amount of injury will rapidly decrease as the pollutant is diluted in the atmosphere either vertically or horizontally. Atmospheric conditions may favor or inhibit the dilution of a pollutant and thereby alter the injury it causes to a plant. In addition, precipitation may react chemically with air pollutants and modify their toxicity to plant tissues.

Wind Direction

Wind movements are critical in understanding where the air pollutants will travel from sources and, therefore, in predicting where injury to plants will most likely occur. Prevailing wind direction varies with time of year and also may be affected by local geographic features such as mountains or bodies of water. The most severe effects of pollutants would be expected downwind from known sources. The wind direction during a brief period of emission, such as a gas leak, may be essential in establishing the cause of the injury. Records of local weather stations can be useful in providing wind direction and other meteorological information relevant to air pollutant movements.

Temperature

Air temperature has two effects on air pollutants: (1) it determines the vertical movement of air pollutants, and (2) it modifies the rate at which pollutants react with plant tissue.

Gases will move from warm to cool air and will normally rise and expand in the atmosphere. Air normally becomes colder with increasing altitude at a rate of approximately 30°F/mile (10°C/km). Therefore, most air containing pollutants will usually rise from the earth's surface into the atmosphere and become diluted before the pollutants contact plants unless they are blown into contact by horizontal winds. These air pollutants, however, will eventually return to earth and come in contact with the soil, vegetation, or some body of water. The earth is the ultimate sink of most air pollutants.

The temperature structure of the atmosphere is not always favorable to gas rising. Sometimes layers of warm air, or inversions, form in the lower atmosphere and air pollutants are not able to rise and disperse in the upper atmosphere. (Fig. 22.1). This restriction of movement can result in prolonged exposure of plants to high levels of phytotoxic air pollutant. Thermal inversions aloft often occur in narrow valleys, close to mountain ranges, and adjacent to large bodies of water. In these locations air near the ground often remains cool while air aloft becomes warmer. Thermal inversions can also occur at the ground level during cold weather when temperatures near the ground are lower than temperatures aloft.

Air pollutants react chemically with the tissues of the leaf. The rate of chemical reactions is governed to a large extent by temperature. Faster chemical reaction rates usually occur at higher temperatures. Gas exchange through stomates is also accelerated at high temperature. Photochemical oxidants are also produced at higher rates with increasing temperatures. Injury from air pollutants, there-

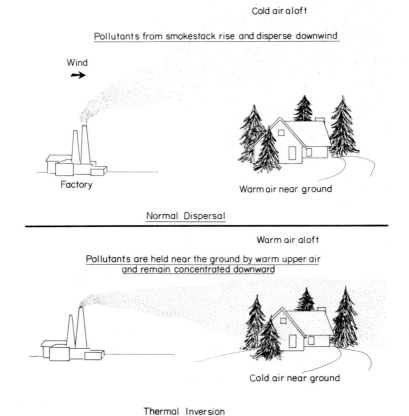

Fig. 22.1 Dispersal of air pollutants during normal thermal conditions in the atmosphere and during a thermal inversion.

fore, is often more severe during sunny hot weather and less severe during cool cloudy weather.

Moisture

Atmospheric Moisture Moisture in the air affects the behavior of a number of particular pollutants. Small particles are often hydroscopic and will absorb moisture from the atmosphere. In some cases the particles go into solution or deliquesce to become liquid droplets. The chemical reaction with water vapor often causes the materials to become more phytotoxic. High atmospheric moisture favors all these reactions. In addition, stomates are usually open during high relative humidity and closed during low relative humidity. High relative humidity, therefore, can often cause increased plant injury from particulate pollutants.

Precipitation Some air pollutants react with moisture either in the form of precipitation or dew and become caustic acids, which are toxic to plant tissues. Rain that has reacted with such pollutants is termed "acid precipitation" (see section later in this chapter). In the case of aerosols and particulates rain may have a beneficial effect on plants by washing pollutants from foliage.

Solar Radiation

Energy from sunlight is critical for the formation of photochemical oxidants. These materials will form at much reduced rates on cloudy days and will not form in the dark. The amount of oxidants, and therefore the amount of plant injury, is determined by the amount of solar energy.

PHYTOTOXIC GASES

Gaseous air pollutants, whether from point or diffuse sources, cause injury to plants by entering the leaf through stomata and causing the injury or death of tissues. However, each phytotoxic air pollutant usually causes a different pattern of injury on the leaves. The major gaseous air pollutants injurious to woody plants will be discussed in the following paragraphs.

Point Surface Pollutants

Sulfur Dioxide Sulfur dioxide (SO_2) is produced primarily by burning of fossil fuels (coal and oil) and smelting of sulfur-containing ores such as those that yield copper, lead, zinc, and nickel. Sulfur dioxide injury, therefore, is found most commonly near coal- or oil-burning electric generating plants and near ore-smelting operations.

Sulfur dioxide injury often results in tip burn in conifer needles and interveinal necrosis or chlorosis in broad-leaved evergreens and deciduous trees. Injury on conifers may also occur as yellow or necrotic bands on the needles in a fascicle. Chronic symptoms from prolonged exposures are often what cause a sparse crown with short, yellow-green tufts of current needles. Chronic symptoms on hardwoods are small leaves, early fall coloration, sprouting, and progressive twig and branch dieback resulting in a sparse crown. Chronic symptoms for sulfur dioxide injury and most other phytotoxic air pollutants may be confused with numerous infectious and noninfectious root diseases. Accurate diagnosis of chronic injury usually requires some determination of air pollutant concentration in the air around trees. At times, sulfur from sulfur dioxide remains in the leaf, and tissue analysis may enable detection of this air pollutant.

TABLE 22.1
Tolerance of Some Woody Plants to Sulfur Dioxide[a]

Tolerant	Intermediate	Sensitive
Arborvitae	Alder, mountain	Alder, thinleaf
Cedar, Western red	Basswood	Aspen
Fir, white	Boxelder	Ash, green
Ginko	Cottonwood	Birch
Hawthorn, black	Dogwood, red osier	Elm, Chinese
Juniper	Douglas fir	Larch, western
Linden, Littleleaf	Elm, American	Maple, Manitoba
Maple, Norway	Fir, balsam	Maple, Rocky Mountain
Maple, silver	Fir, grand	Mulberry, Texas
Maple, sugar	Hawthorn, red	Pine, eastern white
Oak, pin	Hemlock, western	Pine, jack
Oak, red	Honeysuckle, tartarian	Pine, red
Pine, limber	Lilac	Poplar, Lombardy
Pine, piñon	Maple, red	Serviceberry
Poplar, Carolina	Mountain-ash, European	Willow, black
Spruce, blue	Mountain-laurel	
Yew, pacific	Oak, white	
	Pine, Austrian	
	Pine, ponderosa	
	Pine, western white	
	Poplar, balsam	
	Spruce, Engleman	
	Spruce, white	

[a] From Davis and Wilhour (1976).

Sulfur dioxide injury is best controlled by reduction of emission from its source. Since this is not usually feasible without some type of action by government regulatory agencies, most injury from sulfur dioxide and other air pollutants on shade trees is controlled by keeping woody plants vigorous and is avoided by selection of resistant species (Table 22.1). Cotrufo and Berry (1970) found that applications of balanced fertilizer decreased injury to white pine exposed to sulfur dioxide.

Fluorides Fluorine is found in small concentrations in most soils but is tightly bound in various nonmobile forms of fluoride, and as such is nontoxic to plants. Minerals or ores containing fluorides are commonly used in the manufacture of glass, cement, brick, tile, pottery, aluminum, and phosphate fertilizer. When fluorine-containing materials are processed by heating to remove impurities hydrogen fluoride (HF), a gas toxic to plants, is produced.

Injury from hydrogen fluoride appears as tip necrosis in conifers and tip and marginal necrosis of broad-leaved trees (Table 22.2). Red-brown necrosis is considered typical of fluoride injury. Injury in conifers usually begins with yellowing of needle tissues, which progressively turn to tan then red-brown. Injury in

TABLE 22.2
Tolerance of Some Woody Plants to Hydrogen Fluoride[a]

Tolerant	Intermediate	Sensitive
Alder, European black	Arborvitae	Apricot, flowering
Ash, American mountain	Ash, European	Boxelder
Ash, European mountain	Ash, green	Fir, Douglas
Ash, Modesto	Beech, European	Larch, western
Birch, European cut-leaf	Birch, European white	Paulownia
Cherry, Oriental	Chestnut, Spanish	Pine, eastern white
Elder, European	Filbert, European	Pine, loblolly
Elm, American	Holly, English	Pine, Mugho
Juniper	Linden, European	Pine, ponderosa
Linden, American	Locust, black	Pine, Scots
Linden, little-leaf	Maple, hedge	Spruce, blue
Planetree	Maple, silver	
Plum, flowering	Mulberry, red	
Russian olive	Oak, English	
Spruce, white	Planetree, Oriental	
Tree of Heaven	Poplar, Eugene	
Willow	Poplar, Lombardy	
	Walnut, black	
	Walnut, English	

[a] From U.S. Forest Service (1973).

broad-leaved trees usually begins with fading of leaf tissue followed by red-brown necrosis, which is usually sharply defined from the healthy tissue. Emerging leaf tissues are most susceptible to acute injury and consequently most severe injury occurs in the spring. Chronic injury, however, occurs progressively during the entire growing season.

Ethylene   Ethylene is a well-known growth regulator in plants but can also cause injury at high concentrations. Ethylene is a by-product of natural gas processing, polyethylene manufacture, and burning of numerous organic materials such as agricultural waste and residential garbage. Ethylene is also produced during incomplete combustion of fuel by greenhouse heaters and has caused considerable injury to nursery stock. Ethylene injury is similar to that produced by plant hormone-type herbicides—e.g. growth reduction, leaf and bud abscission, leaf yellowing, and necrosis and flower deformities (see the section on herbicide injury in Chapter 20, Chemical Injury). Ethylene injury, therefore, is sometimes confused with that from herbicides.

Ammonia   Ammonia is a commonly used chemical around the home that often gets into the air at low levels from burning of home fuels and from automobile exhausts. Injury to plants, however, is most often associated with industrial spills such as breaks in pipelines or accidents involving rail cars, tank trucks, or fertilizer (anhydrous ammonia) tanks that carry ammonia. Ammonia injury is quite variable and dependent on plant species. Symptoms of injury reported on spruce are red-yellow younger needles and black older needles. Broad-leaved trees often display interveinal necrosis, which often resembles sulfur dioxide injury, and marginal chlorosis.

Chlorine and Hydrogen Chloride   Chlorine, like ammonia, is usually associated with plant injury following industrial spills, such as leaks from storage tanks or pipelines containing chlorine, from swimming pool chlorination systems, and from the manufacture of chlorine and sodium hypochlorite. Chlorine is also produced as a by-product of petroleum refining and the manufacture of flour, glass, pulp, and textiles. Hydrogen chloride is most often produced during the incineration of polyvinyl chloride (PVC) plastics. These plastics are commonly used and are important in the incineration of domestic waste. Injury from both chlorine and hydrogen chloride includes chlorosis, stipple (many small spots), necrosis, and reddening. These chemicals can react with water during light rain or fog resulting in an acid mist that can cause necrotic spots.

Diffuse Oxidants

Oxidants are not emitted from a single source but occur over a large area from many original sources. Oxidants are gases, principally ozone, that are produced

TABLE 22.3
Tolerance of Some Woody Plants to Ozone[a]

Tolerant	Intermediate	Sensitive
Arborvitae	Boxelder	Ash, green
Birch, European white	Cedar, incense	Ash, white
Dogwood, white	Cherry, Lambert	Aspen, quaking
Fir, balsam	Elm, Chinese	Azalea
Fir, Douglas	Gum, sweet	Cotoneaster
Fir, White	Larch, Japanese	Honey locust
Gum, black	Lilac	Larch, European
Holly	Oak, black	Mountain-ash, European
Linden, American	Oak, pin	Oak, Gambel
Linden, little-leaf	Oak, scarlet	Oak, white
Maple, Norway	Pine, eastern white	Pine, Austrian
Maple, sugar	Pine, lodgepole	Pine, Jack
Oak, English	Pine, pitch	Pine, Jeffrey
Oak, red	Pine, Scotch	Pine, loblolly
Pine, red	Pine, shortleaf	Pine, Monterey
Spruce, blue	Pine, slash	Pine, ponderosa
Spruce, Norway	Pine, sugar	Pine, Virginia
Spruce, White	Redbud, eastern	Poplar, tulip
Walnut, black		Sycamore, American
Yew		Tree of Heaven
		Walnut, English

[a] From Davis and Wilhour (1976).

in the atmosphere by chemical reactions with oxygen. Ozone is a natural component of the atmosphere and is always present in the air at low levels. In the
upper stratosphere natural ozone serves a beneficial purpose of screening out
dangerous high energy radiation from the sun and outer space. Small amounts
of ozone are also produced by lightning. These natural sources, however, cause
very little injury to plants while high amounts of ozone associated with the by-
products of motor vehicles can cause considerable injury to plants. Automobiles,
trucks, and other vehicles with internal combustion engines emit oxides of nitrogen and hydrocarbons in their exhausts. Power plants that burn fossil fuels are
also a source of oxides of nitrogen. These chemicals react in the atmosphere with
the aid of sunlight with oxygen to produce ozone, peroxyacetylnitrate (PAN),
and numerous other compounds. Both ozone and PAN are toxic to plants but
ozone is much more abundant and injurious to woody plants.

Ozone injury appears to be widespread on both conifers and deciduous trees
in rural areas as well as in urban and suburban areas (Table 22.3). Typical
symptoms of injury on conifers include chlorotic mottle (small patches of yellow
or brown tissue mixed with the green healthy tissues), yellow or necrotic ban-
ding, tip burn, chlorotic dwarfing (yellow-green foliage on stunted trees), and a
tufted appearance due to early defoliation of older needles. Broad-leaved trees
often exhibit as symptoms of ozone injury stippling, or red-purple or necrotic
flecks on the upper leaf surfaces.

PARTICULATES AND AEROSOLS

Air pollutants can occur as small solid particles, fine liquid mists, or aerosols
that can remain suspended in the atmosphere for limited periods. Both solid
particles and liquid droplets, however, will combine with other particles or
droplets and eventually fall back to earth. When these materials contact plant
foliage, injury can occur from both physical and chemical interactions with the
tissues.

Particulate air pollutants can occur as soot or flyash from burning or as dusts
from numerous industrial operations such as cement manufacture and stone
crushing. Most injury to plants occurs from dusts, which can reduce photosynthesis by coating the foliage, react chemically with the leaf tissues, or plug
stomata. Dusts may also affect root growth by altering soil conditions through
long-term buildup. Rhoads (1976) reported that long-term deposition of cement
dust on soil beneath oak trees resulted in severe foliar injury and decline.
Buildup of heavy metals to toxic concentration in soil can also occur from prolonged deposition of certain dusts around trees.

Aerosols can also be toxic to trees. Most phytotoxic aerosols are caustic acids,
such as sulfuric acid or hydrochloric acid, which are sometimes emitted during
their manufacture or use. Fine mists may be generated directly at the source or
may occur in the atmosphere when particulates combine with fog and mist.

When the aerosol droplets contact foliage, local injury to the upper leaf surface commonly occurs. In severe cases injury extends through the leaf and necrotic tissue may drop out leaving a "shot hole" effect.

ACIDIC PRECIPITATION

It has been known for many years that air pollutants can react with precipitation and change its chemical properties. Many pollutants associated with human activities, such as sulfur dioxide, oxides of nitrogen, fluorides, and chlorides, react with precipitation to produce acids. These pollutants have caused a lowering of the pH of rain and snowmelt in large areas of the world, often in regions quite distant from industrial centers. This low-pH rain is termed *acid rain* or *acidic precipitation*. The effects of acid rain on the pH of certain lakes and streams and its environmental consequences have been a source of concern for many environmental scientists.

The general public often associates acid rain with all injurious effects of air pollutants on trees. However, air pollution is actually very complex. Airborne chemical deposition occurs all over the world, mostly in solid "dry" form, from emission sources that are also widely scattered. Popular articles have blamed acid rain for the decline of both forest and shade trees all over the world. Despite extensive research by many environmental scientists, there have not been any widely accepted studies that have attributed the decline of forest or shade trees to the effects of acid rain alone.

MIMICKING SYMPTOMS

The symptoms of air pollution injury on plants are not unique but are also produced by a number of noninfectious diseases and even some infectious diseases. These diseases mimic the symptoms of air pollution injury and make accurate diagnosis by symptoms extremely difficult. They can cause yellowing, tip and marginal necrosis, scorch, and leaf spots. Some of the most common disease problems that mimic air pollution injury are nutrient disorders, moisture and temperature stress, virus disease, mycoplasma diseases, and some leaf spots caused by bacteria and fungi. In addition, insects with piercing–sucking mouth parts, such as aphids and leafhoppers, and mites can cause a stippling or mottling on leaves, from their feeding in high populations, which is very similar to oxidant injury. In diagnosis of air pollution injury one must first eliminate these common mimicking disease and insect problems before considering air pollutants as the primary cause of plant injury.

LITERATURE CITED

Cotrufo, C., and C. R. Berry. (1970). Some effects of a soluble NPK fertilizer on sensitivity of eastern white pine to injury from SO_2 air pollutants. *For. Sci.* **16,** 72–73.

Davis, D. D., and R. G. Wilhour. (1976). Susceptibility of woody plants to sulfur dioxide and photochemical oxidants. *Environ. Prot. Agency (U.S.), Rep.* **600/3-76-102,** 1–72.

Rhoads, A. F. (1976). Forest species show a delayed response to cement dust in the soil. *J. Arbori.* **2,** 197–199.

U.S. Forest Service. (1973). Trees for polluted air. *U.S. Dep. Agric., Misc. Publ.* **1230,** 1–11.

SUGGESTED REFERENCES

Anonymous. (1974). Urban plants vs. pollution. *Agric. Res.* **23,** 3–6.

Brennan, E., and A. F. Rhoads. (1976). The response of woody species to air pollutants in an urban environment. *J. Arbori.* **2,** 1–5.

Campana, R. J. (1976). Air pollution effects on trees. *Trees Mag.* **35,** 35–38.

Davis, D. D., and H. D. Gerhold. (1976). Selection of trees for tolerance to air pollutants in Better trees for metropolitan landscapes symposium proceedings. *U.S., For. Serv., Gen. Tech. Rep.* **NE-22,** 61–66.

Dochinger, L. S. (1972). Can trees cleanse the air of particulate pollutants? *Proc. Int. Shade Tree Conf.* **48,** 45–48.

Jacobson, J. S., and A. C. Hill (eds.). (1970). Recognition of air pollution injury to vegetation: A pictorial atlas. Air Pollut. Control Assoc., Pittsburgh, Pennsylvania.

LaCasse, N. L., and W. J. Moroz. (1969). Handbook on effects assessment: vegetation damage. Cent. Air Environ. Stud., Pennsylvania State University, University Park.

LaCasse, N. L., and M. Treshow (eds.). (1976). Diagnosing vegetation injury caused by air pollution. *Environ. Prot. Agency (U.S.) Publ.* **68-02-1344.**

Loomis, R. C., and W. H. Padgett. (1975). Air pollution and trees in the east. *U.S., For. Serv.* **S&PF-8,** 1–28.

Roberts, B. R. (1971). Trees as air purifiers. *Proc. Int. Shade Tree Conf.* **47,** 22a–25a.

Skelly, J. M., and J. B. Will. (1974). The use of fertilizer to alleviate air pollution damage to white pine (*Pinus strobus*) Christmas trees. *Plant Dis. Rep.* **58,** 150–154.

Treshow, M. (1970). Environment and plant response. McGraw-Hill, New York.

Wood, F. A. (1967). Air pollution and shade trees. *Proc. Int. Shade Tree Conf.* **43,** 66–82.

23

Diebacks and Declines— Complex Diseases

INTRODUCTION

The diseases of shade trees discussed in previous chapters were caused primarily by the action of one infectious or one noninfectious agent. There, are however, many cases where the cause of the decline or death of trees cannot be attributed to only one agent. In these cases primary stress or stresses cause(s) the health of the tree to deteriorate and any number of secondary pathogens or insects may cause the eventual decline and death of the tree. These diseases are usually called diebacks or declines, or more simply complex diseases. Since a number of stresses, such as construction injury, moisture imbalance, soil compaction, chemical injury, or insect defoliation, can cause the deterioration of a tree's health and since a large number of secondary pathogens and/or insects may attack the weakened tree, there exists a great deal of confusion concerning complex diseases. The concepts and examples discussed below are those developed and presented by David R. Houston and his co-workers Johnson Parker and Philip M. Wargo of the U.S. Forest Service Laboratory in Hamden, Connecticut.

Complex diseases can be understood best by comparing them to diseases with single primary causes such as Dutch elm disease or white pine blister rust (Fig. 23.1). In these latter diseases the condition of the host and the environment play only a relatively minor role in the disease. Thus, the primary pathogen can attack vigorous, healthy trees. In dieback and decline diseases, however, the host condition as altered by adverse environmental factors is of major importance. The organisms of secondary action are able to succeed only when trees have been altered by stress. Often, however, it is the organisms of secondary action that are thought to be wholly responsible for the disease. In summary, dieback and decline diseases are caused by the successive action of primary stress factor(s) followed by organisms of secondary action that can successfully attack only weakened trees.

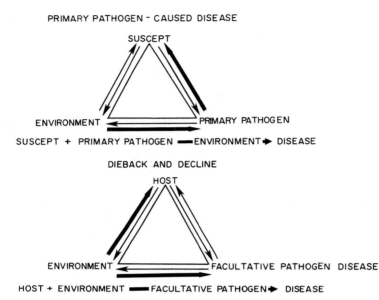

PRIMARY PATHOGEN - CAUSED DISEASE

SUSCEPT + PRIMARY PATHOGEN ➡ENVIRONMENT➡ DISEASE

DIEBACK AND DECLINE

HOST + ENVIRONMENT ➡FACULTATIVE PATHOGEN➡ DISEASE

Fig. 23.1 Comparison of primary pathogen-caused diseases and dieback and decline diseases. (Drawing courtesy of David R. Houston, USDA Forest Service, Hamden, Connecticut.)

Diebacks and declines are common diseases of roadside urban and suburban shade trees due to the magnitude and multiplicity of stresses that occur on trees in these locations. Disease complexes also pose problems for both the plant pathologist and the professional arborist because the initiating stress is often obscure or no longer exists. Homeowners familiar with disease and insect problems with simple controls may also be unsympathetic and hesitant to accept the complex cause–effect relationships of dieback–declines.

An examination of several representative dieback and decline diseases will enable a better understanding of the complex nature of these diseases. In the following sections, three diseases—maple decline, oak decline and mortality, and ash dieback—will be discussed in detail.

MAPLE DECLINE

Sugar maples have been planted as roadside shade trees in large numbers since colonial times. More recently this species was chosen in many cases to replace the large numbers of elms lost along streets, around homes, and in recreation areas due to the Dutch elm disease. A survey conducted by the Shade Tree Laboratories of the University of Massachusetts found 80 to 90% of newly planted trees in the 1950s were maples. Sugar maple and other maple species were chosen primarily because they were readily available at nurseries and were inexpensive. Sugar maple, in contrast to American elm, had not been proved to be an adaptable tree species for suburban and urban plantings. The natural range of sugar maple is restricted to colder climates in the northeast and north central United States, and this species is intolerant to warmer temperatures. Sugar maple grows best in deep fertile soils with abundant moisture, conditions rarely found along streets and in most locations where shade trees are planted. Also sugar maple is relatively intolerant of most stresses associated with activities of people such as site changes, root injuries, deicing salt, poor soil condition, soil moisture extremes, and frequent wounding.

Maple decline is primarily the failure of sugar maples to be grown successfully as shade trees in many locations. Although maple decline occurs occasionally on other widely planted species including Norway, red, silver, and sycamore maples, sugar maples suffer the most. Since this species is known to be intolerant of many of the activities of people and of site and environmental conditions that shade trees often are exposed to, many living sugar maple shade trees have been under stress for many years. It is the magnitude, frequency, and timing of stress, however, that determine both the response of the tree and the likelihood of attack by organisms of secondary action.

The responses of maple trees to stress can often be detected in the early stages (Fig. 23.2). Some of the earliest symptoms are early fall coloration, late spring flush, decrease in internodal twig growth, and progressive decrease in tree ring thickness. As stress continues, foliage may be noticeably smaller, lighter green, and produced in "tufts" or "clumps" on sprout-origin tissue at nodes. The entire tree may appear thinner. Twig dieback occurs initially in the upper branches with lateral buds sprouting. Heavy seed crops may occur. Symptoms increase in severity as the tree responds to the continued effects of stress. The progressive death of numerous upper branches (dieback) occurs. Dieback may not necessarily be detrimental to the tree's health but may be a survival mechanism to allow the tree to decrease the amount of living tissues that must be maintained during times of stress. On the other hand, dieback is a symptom of severe stress and of a set of conditions that may lead to the eventual death of the tree if the stress is not relieved.

The symptoms of maple decline and other complex diseases are the result of the inability of the root system to supply the water, nutrients, and energy reserves that are needed by the aboveground parts of the tree. It is, therefore, not surprising to find similar "dieback and decline" symptoms resulting from root disorders of both infectious and noninfectious origin. A disruption of the balance between roots and branches occurs in both these cases. A restoration of the balance results as a consequence of dieback of branches. The lessened demand for root reserves improves the chances for survival of the tree. Whether or not the tree will survive is related to the magnitude, duration, and nature of the stress and the vigor of the tree.

Severe stress can sometimes cause the death of trees without the attack of

Fig. 23.2 Sugar maples in progressive stages of maple decline. (A) Healthy; (B–E) increasing stages of decline. [Photo (A) courtesy of Shade Tree Laboratories, University of Massachusetts, Amherst. Photos (B–E) courtesy of Dennis Newbanks, University of Massachusetts, Amherst.]

secondary organisms but in many cases chronic stress lowers the defense mechanisms of trees to allow such an attack. In sugar maple numerous secondary organisms may attack a stressed tree but the shoestring fungus, *Armillaria mellea*, is the most common. It has been found (Parker and Houston 1971; Wargo 1972) that stressed sugar maple trees mobilize their energy reserves from starch to simple sugars, a change that favors the growth of *A. mellea*. Organisms of secondary action, such as *A. mellea*, can most readily attack sugar maples and other trees that have altered their metabolism in response to stress (Wargo and Houston, 1974).

Maple decline can be prevented, in most cases, by eliminating stress on trees or by not planting sugar maples where stress is unavoidable. These recommendations, however, are often difficult to follow because large numbers of sugar maples already are growing in locations where stress is unavoidable. Maple decline can be treated most successfully in its early stages. Early recognition and alleviation of stress are the keys to controlling the diebacks and declines. But once the disease has begun, steps must be taken to restore the balance between root and shoots and to stimulate root growth. Control measures, therefore, are similar to those for alleviating other root stress problems, such as transplanting, cut roots, raise of grade, underground gas, and flooding. Remove the excess branches that the roots cannot support by pruning all dead and dying branches and 10 to 20% of the live branches. Stimulate root growth by fertilizing with compounds high in phosphorus and low in nitrogen, such as phosphate or superphosphate fertilizer, and by watering at regular intervals throughout the summer and into late fall. Thick mulches should also be placed over as much as the root system as possible to create a favorable environment for root growth. High nitrogen fertilizers should be avoided because the stimulation of excess branching and foliage may only aggravate the stress on the root system.

OAK DECLINE AND MORTALITY

There are a large number of oak species used as shade trees throughout North America. Some are forest trees that became shade trees when homes were built on wooded lots, but most were planted. Most oaks grow best on acid soils or on soil acidified by the accumulation of their fallen leaves over many years. Oaks are not tolerant of natural alkaline soils or heavily limed soils. Oaks, like most other trees, prefer deep fertile soil with abundant moisture. The conditions that favor the best growth of oaks, however, are not present in many urban locations.

Oaks are subjected to many stresses that lower their vigor. Soil conditions that favor most oak species are not those that favor growth of turfgrass, the major object of attention by the homeowner. The slightly alkaline soil best suited for growth of turfgrass does not favor oaks, and applications of lime-based fertilizers often adversely affect oaks planted on lawns. Moisture is often limited in soil around shade trees due to excessive drainage and competition with turfgrasses. Soil fertility along roadsides is also generally too low for shade trees. Oaks are often defoliated by such insects as the gypsy moth, elm spanworm, oak leaf rollers, leaf tiers, canker worms, and many others. Oaks are also quite intolerant of site disturbances associated with the activities of people. Many native oaks, therefore, are unable to survive the changes associated with urbanization.

The responses of oaks to primary stress are similar to those of maple and most other woody plants. Early symptoms include premature fall coloration, late spring flush, and decreased twig and stem growth. Increased stress effects include chlorosis, twig dieback, sprouting on branches and trunk, tufted foliage, general thinning of foliage, and heavy acorn production. Dieback of larger branches increases progressively with stress and in severe cases most of the foliage occurs as sprouts on the trunk and main leaders.

Numerous secondary organisms may attack stressed oaks, just as they attack stressed trees of any species, but the shoestring fungus, *A. mellea,* and the two-lined chestnut borer, *Agrilus bilineatus,* are the most common organisms of secondary action. Houston (1973) has reported that oaks under stress undergo changes in carbohydrate metabolism similar to those that favor attack of sugar maple by *A. mellea.*

Oak decline can be prevented in most cases by providing adequate soil moisture, keeping soil pH mildly acidic, and by avoiding site disturbances around trees (Ware and Howe, 1974). Oak decline in the early stages can be controlled by relieving the primary stress and by helping the tree to restore a normal balance between the roots and branches. Dead and dying branches should be removed along with 10 to 20% of the live branches. Watering should be done at regular intervals into the late fall and high-phosphate fertilizers low in nitrogen should be applied to stimulate root growth. Mulches over the root system should be added to a depth of at least 4 inches (10 cm). But stonechip mulches containing limestone should not be used around oaks.

ASH DIEBACK

White ash, and to a lesser extent green ash, have been planted extensively as shade trees. These fast-growing trees are often intolerant of the limited moisture conditions found along the roadside and in yards. Established ash trees are also intolerant of site disturbances that can result in decreased water availability to the roots. They often outgrow their limited sites and suffer from lack of root space. Since these trees are often drought-sensitive it is not surprising that ash dieback is most severe during periods of limited moisture. White ash is also known to be sensitive to ozone, a common pollutant in most urban and suburban areas.

Ash responds to primary stress in the early stages by displaying premature fall coloration, late spring flush, and decreased twig and ring diameter growth. As stress continues foliage often turns yellow-green, twigs die back, and adventitious buds at nodes produce shoots that result in tufts of chlorotic foliage below the dead twigs. The foliage on the entire tree may appear thin and heavy seed crops can occur. With continued stress sucker growth may also appear on larger branches as branches die back in the upper portions of the tree.

Stressed ash trees are often susceptible to the attack of weak canker fungi, especially *Cytophoma pruinosa* and a *Fusicoccum* sp. These fungi, which accelerate the dieback by girdling branches and main stems, are known to be normal inhabitants of the bark on lower shade-weakened branches. White ash is also known to be a host to several viruses and to the ash yellows mycoplasma, but the role of these agents in ash decline is as yet unknown.

OTHER COMPLEX DISEASES

While all trees are susceptible to dieback and decline as a result of the combined effects of primary stress and attack of secondary organisms, some other notable examples of similar complex diseases deserve mention here. Beech bark disease and birch dieback are two complex diseases that have resulted in major losses to forest trees. In the beech bark disease, the American beech is stressed by the attack of the beech scale insect, *Cryptococcus fagi,* followed by attacks of the canker fungus, *Nectria coccinea* var. *faginata.* The fungus causes extensive killing of bark on the main stem (Fig. 23.3). Most trees die from further secondary invasion by woodrot fungi on the stem and root rot fungi like *A. mellea,* and many are killed by the initial cankers. Birch dieback is felt to be the result of a combination of numerous adverse stress factors, such as excessive thinning, drought, extremes of temperature, and attacks of the bronze birch borer, as well as defoliating, leaf-mining, and sucking insects. Affected trees initially exhibit twig dieback and sucker growth on the main branches and trunk. The bronze birch borer frequently attacks these weakened trees. Dieback progresses to larger branches resulting in the eventual death of the trees. Secondary fungi such as *A. mellea* are frequently found on roots of dying trees.

CONCLUSIONS

No species of woody plant is immune to the effects of complex diseases but some commonly planted tree species appear to be better able than others to tolerate many of the stresses of the urban environment. Diebacks and decline diseases can be minimized by using species known to tolerate stresses associated with the shade tree environment and the activities of people. Planting a variety of species will also help reduce the impacts of diebacks and declines. Trees should be selected for their tolerance to stress. The selection of smaller and slower-growing trees that can adapt more effectively to stress and demand less of their site will also decrease the incidence of diebacks and declines.

Diebacks and declines can often be controlled by restoring the balance between roots and branches through pruning, stimulating root growth through fertilization with compounds high in phosphorus and low in nitrogen, by regular watering, and by adding thick mulches. Early detection of stress is the key to success in controlling decline diseases since the ability of the tree to respond to therapeutic treatments depends upon its vigor and in the turning on of its

Fig. 23.3 Beech bark disease on American beech. Bark killing associated with beech scale insect–canker fungus complex. (Photos courtesy of Alex L. Shigo, USDA Forest Service, Durham, New Hampshire.)

defense mechanisms. Better methods of early detection are needed in all tree diseases, but they are especially needed for diebacks and declines where the onset of easily visible symptoms is often an indication of severe and chronic stress (see Chapter 24, Nonpathogenic Conditions).

LITERATURE CITED

Houston, D. R. (1973). Dieback and declines: Diseases initiated by stress, including defoliation. *Proc. Int. Shade Tree Conf.* **49**, 73–76.

Parker, J., and D. R. Houston. (1971). Effects of repeated defoliation on root and root collar extractions of sugar maple trees. *For. Sci.* **17**, 91–95.

Ware, G., and V. K. Howe. (1974). The care and management of native oaks in northern Illinois. *Morton Arb. Plant Inf. Bull.* **4**, 1–4.

Wargo, P. M. (1972). Defoliation induced chemical changes in sugar maple roots stimulate growth of *Armillaria mellea*. *Phytopathology* **62**, 1278–1283.

Wargo, P. M., and D. R. Houston. (1974). Infection of defoliated sugar maple trees by *Armillaria mellea*. *Phytopathology* **64**, 817–822.

SUGGESTED REFERENCES

Brandt, R. W. (1963). Ash dieback in New England and New York. *Proc. Int. Shade Tree Conf.* **39**, 38–43.

Fisher, R. (1973). Trees grow up by growing down. *Proc. Int. Shade Tree Conf.* **49**, 85–87.

Halliwell, R. S. (1964). Live oak decline. *Proc. Int. Shade Tree Conf.* **40**, 178–180.

Houston, D. R. (1967). The dieback and decline of north eastern hardwoods. *Trees Mag.* **28**, 12–14.

Houston, D. R. (1971). Noninfectious diseases of oaks. *Proc. Oak Symp. U.S. For. Serv.*, pp. 118–123.

Houston, D. R. (1974). Diagnosing and preventing diebacks and declines of urban trees: Lessons from some forest counterparts. *Morton Arb. Q.* **10**, 55–59.

Howe, V. K. (1974). Site changes and root damage: Some problems with oaks. *Morton Arb. Q.* **10**, 49–53.

Mader, D. L., B. W. Thompson, and J. P. Wells. (1969). Influence of nitrogen on sugar maple decline. *Mass., Agric. Exp. Stn., Bull.* **582**, 1–24.

Ross, E. W. (1966). Ash dieback etiological and developmental studies. State Univ. Coll. *For. Syracuse Univ., Tech. Publ.* **88**, 1–80.

Shigo, A. L. (1976). The beech bark disease. *J. Arbori.* **2**, 21–25.

Sinclair, W. A. (1966). Decline of hardwoods? Possible causes. *Proc. Int. Shade Tree Conf.* **42**, 17–32.

Skelly, J. M. (1974). Growth loss of scarlet oak due to oak decline in Virginia. *Plant Dis. Rep.* **58**, 396–399.

Wargo, P. M., J. Parker, and D. R. Houston. (1972). Starch content in roots of defoliated sugar maple. *For. Sci.* **18**, 203–204.

Welch, D. S. (1963). Maple decline in the northeast. *Proc. Int. Shade Tree Conf.* **39**, 43–48.

PART III

SPECIAL TOPICS

24

Nonpathogenic Conditions

INTRODUCTION

Trees are complex organisms that can confuse the concerned but untrained observer. Any normal but rarely occurring noticeable change in appearance of the tree may be considered a disease or insect problem by the homeowner. Exotic trees challenge both homeowner and arborist. Most nonpathogenic conditions that arouse public concern can be associated either with a misunderstanding about trees or with the presence of nonparasitic organisms that cause little or no harm to the tree. These conditions, however, can be a problem since the homeowner may fear that they threaten the tree or may feel that they detract from the attractiveness of shade trees. Nonpathogenic conditions are treated primarily through education.

MISINFORMATION ABOUT TREES

Fall Coloration

The leaves of deciduous hardwoods begin to change color in midfall, and the resulting bright color patterns make fall foliage viewing an attraction in many areas. Some leaves on each evergreen tree also turn color each fall. The oldest (innermost) leaves on broad-leaved evergreens and on conifers turn brown. When gradual, the color change in these leaves is often overlooked. In some years, however, the color changes suddenly and the homeowner notices it. In some species, such as eastern white pine, most needles are held for only 2 years. About half of the tree's needles turn yellow in the fall. From a distance the entire tree may appear deathly ill, but soon the needles turn brown and are shed, leaving a green, healthy looking tree again. This entire process is normal. Normal fall color changes should occur on all the trees of the same species in a particular area at about the same time and should occur only on the innermost (oldest) leaves while the outer (younger) leaves remain green.

Yellow Varieties of Trees and Shrubs

Leaves of plants are normally green, and yellow foliage is usually considered a symptom of plant disease. However, plants with yellow leaves or with patches of yellow or white tissue among the green leaf tissue (variegated) are often perfectly healthy plants. Many trees and shrubs have been selected for foliage with total or partial lack of green leaf color. Some species, like the Vicary Golden privet, remain yellow all season, while others, like the Sunburst honey locust, are yellow while the foliage is immature but turn green as the leaves age. All these plants are healthy but have undergone some type of mutation that prevents normal green color from forming in all leaf tissue. The "abnormally colored" woody plants are often valuable additions to a landscaped area; a brief description of such use is presented by Rothenberger (1976).

Fall Leaf Abscission

Leaf abscission follows fall coloration in most deciduous hardwoods. Some, however, such as beech, chestnut, and many oaks, retain many of the dead leaves throughout the winter. These leaves are shed in the spring as the new foliage is produced. Leaf abscission also follows the normal fall coloration in most evergreens (Fig. 24.1). Again, however, some broad-leaved evergreens, such as rhododendrons, hold their brown leaves through the winter. Fall abscission of older foliage, like fall coloration, is normal.

Twig Abscission

A specialized layer of cells sometimes forms at the base of twigs that allows the tree to drop excess twigs, in a manner similar to normal shedding of dead foliage in the fall. The branches of a tree are composed of elongating (terminal) branches and secondary (lateral) branches. The elongating branches continue to grow each year as the crown of the tree expands. Some of the secondary branches can become elongating branches, but most become shaded out and are cast. If abscission did not occur, the interior branches would be covered with a tangle of old dead twigs. Homeowners, however, may be startled to find twigs on the lawn during the summer. This condition is sometimes confused with the attack of twig-feeding insects or animal injury. Normal twig abscission occurs only at the base of the twig or at the "terminal bud scale scars" and makes a clean, smooth break. Insect or animal feeding cause a jagged or torn edge or leaves a semicircular depression from the larval gallery. Teethmarks are sometimes found on twigs cut by animals.

Shedding or Cracking of Outer Bark

The appearance of the outer bark changes over the life of a tree and can be a source of concern. The outer bark of a woody plant cracks and may eventually be

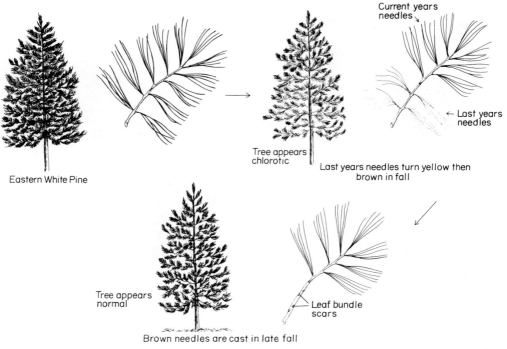

Fig. 24.1   Fall needle drop of conifers.

shed as the trunk and branches increase in diameter. Most young trees have a relatively smooth bark. Depending on species, the outer bark may develop any of a variety of peeling and cracking patterns as the tree grows. Color changes may also accompany the outer bark splits; for example, white and gray birches both have dark brown outer bark until sufficient splitting occurs to reveal the white bark beneath. Some trees, such as shagbark hickory and sycamore, have loose outer bark that is conspicuous and is frequently shed (Fig. 24.2). All these processes are perfectly normal and are essential to allow the tree to grow.

Flower, Fruit, and Seed Production

Flowers are the reproductive structures of all trees and shrubs. Many species are planted primarily for the flowers they produce, such as azalea, crabapple, dogwood, forsythia, golden-rain, lilac, and rhododendron. Fruits are also attractive on some species such as barberry, crabapple, dogwood, and holly, and seeds often benefit wildlife. Sometimes, however, fruits and seeds are considered a nuisance by homeowners because they collect on lawns, sidewalks, and driveways.

Flowering in trees is not automatic each year and is controlled by numerous

Fig. 24.2   Normal bark peeling and shedding on an American sycamore (A). Close-up (B).

physiological and environmental factors. Flower production does not begin until a tree reaches physiological maturity, which may range from a few to over 20 years. Shock from transplanting usually prevents flower production for 3 to 5 years. Extreme cold during winter or just before bud break will often kill flower buds. For example, it is common in the northern United States to find forsythia producing abundant flowers on the lower half of the shrub which was protected by snowcover and none on the exposed upper half. Conditions that favor the best growth of a tree, such as adequate sunlight, soil moisture and nutrition, root space, and moderate temperatures will also favor flower production.

Fruit and seed production can only occur after flowers have been produced and pollinated. Many trees have both male and female parts on the same flower (perfect flowers) and can sometimes be self-pollinated; however, many such tree species (like elm) are not self-fertile. Other trees (like pine) have separate male and female flowers, and still others have only male or female parts (imperfect flowers). These rely solely on insects or wind for pollination. Sometimes only male or only female flower-bearing trees are found in an area; pollination does not occur and no fruit forms. Two examples of trees in this group are holly and ginkgo. Female holly trees produce colorful red berries only when a male holly is also growing nearby. Female ginkgos also produce fruits only when both species are present. Ginkgo fruit, however, has a foul odor when ripe, so planting of male and female ginkgo trees together is avoided. If wet weather persists through the bloom period little wind pollination can take place. Likewise, cold weather during the bloom period may inhibit insect pollination. Then little fruit and seed will be produced.

Bright yellow, wind-carried pine pollen sometimes coats foliage and bark of pine trees and smaller plants beneath pines when wind is slight during pine flowering. This yellow coating sometimes creates fears of some mysterious "yellow disease." Pollen is harmless to the plants and is washed from the foliage by rain.

NONPARASITIC ORGANISMS

Sooty Mold

Aphids are sucking insects that feed on the leaves of both deciduous hardwoods and evergreens. They excrete droplets of a sugary solution called honeydew, which coats the leaves below. Certain fungi grown on the honeydew and produce dark spores that make the leaf surface appear black or sooty. This condition is called sooty mold (Fig. 24.3). The fungi do not penetrate the leaf and

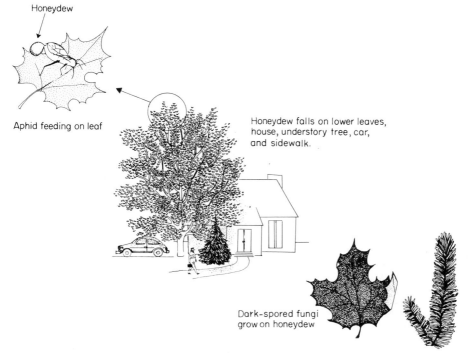

Honeydew

Aphid feeding on leaf

Honeydew falls on lower leaves, house, understory tree, car, and sidewalk.

Dark-spored fungi grow on honeydew

Fig. 24.3 Sooty mold.

can be easily rubbed or washed off. In heavy accumulation, sooty mold can prevent enough light from reaching the leaves, and thus shade out lower branches. The blackened leaves are often noticed by the homeowner and detract from the aesthetic appearance of a tree or shrub. Sooty molds will also form in the honeydew on any object below the area of aphid feeding. Although the leaves on the lower branches of an aphid-infested tree are most commonly affected, honeydew also coats branches, small trees and shrubs, the ground, sidewalks, houses, and any other objects below the tree, such as cars, lawn furniture, and picnic tables. The sticky honeydew is often a nuisance even before it is colonized by the fungi that turn it black. During wet periods it can also become slippery and become a hazard to pedestrians and motorists.

Control of sooty mold is rarely needed to protect shade tree health but is sometimes requested to prevent a nuisance around the home. Sooty mold can be

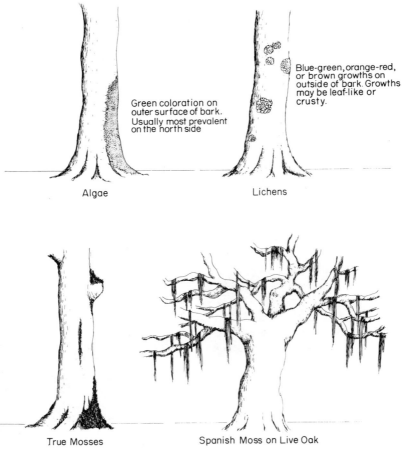

Fig. 24.4 Common epiphytes on trees.

prevented by controlling aphids each year with insecticides. Spraying after honeydew or sooty mold have formed do not cause the problem to disappear, but prevent it from getting worse. Sooty mold, however, will wash off branches and evergreen foliage by the next spring. Objects coated with sooty mold can be cleaned with soap and water, or will eventually "weather" clean.

Algae, Lichens, and Mosses

Many green plants grow on the bark of trees, using the tree only for support. These are called epiphytes and do not harm the tree (Figs. 24.4, 24.5). They produce their own food from H_2O and CO_2 via photosynthesis using sunlight. Common epiphytes on trees are the algae, lichens, and mosses. Although considered attractive by most homeowners, they sometimes are mistaken for a disease.

Algae are primitive green plants that are most abundant in bodies of water but can live wherever moisture is plentiful. The lower trunks of trees in moist, shaded areas are frequently covered with algae, which usually appear as a thin green coating. The only effect algae have on a tree's appearance is the addition of a greenish tint to the bark, so they are not even noticed by many people.

Fig. 24.5 Spanish moss adds to the beauty of a bald cypess (A) and a slash pine (B) at Cypress Gardens, Florida.

Lichens are plants that are a combination of green algae and fungi. These two organisms derive mutual benefits from their association. In a lichen the alga produces food from sunlight while the fungus provides support and absorbs moisture. Lichens form on a variety of moist surfaces in addition to the bark of trees, such as rocks, soil, fallen logs, and old boards. They can appear crusty or leaflike and can occur in a variety of colors, such as brown, gray, green, yellow, and white. Lichens are very sensitive to air pollution and are rarely found near major cities. Their occurrence usually indicates clean air. Most people find lichens attractive but some confuse them with pathogens or diseases. When most homeowners realize lichens do not harm trees and could indicate pure air, concern about lichens ceases.

Mosses are small green plants abundant on moist forest soils, fallen logs, trees, and rocks. Close inspection of mosses will reveal small stems and leaves. Mosses often grow on the lower trunk, on low branches, in bark crevices or in any bark irregularities that give them a place to become established. Mosses, like algae and lichens, cause no harm to the tree and should not be disturbed.

LITERATURE CITED

Rothenberger, R. (1976). Yellow leaves set off trees and shrubs. *J. Arbori.* **2**, 200.

SUGGESTED REFERENCES

Crocket, J. V. (1972). Trees. Time-Life Books, New York.
Ketchum, R. M. (1970). The secret life of the forest. American Heritage, New York.
Kozlowski, T. T. (1962). Tree growth. Ronald Press, New York.
Wilson, B. F. (1970). The growing tree. Univ. of Massachusetts Press, Amherst & Boston.

25

Disease Diagnosis

INTRODUCTION

Diagnosis is the first step in the treatment of disease. Accurate diagnosis of shade tree diseases is a complex activity requiring a combination of many disciplines related to tree health, such as botany, entomology, microbiology, plant pathology, plant physiology, and soil science, as well as broad exposure to tree diseases in the field. Diagnosis also improves with prolonged experience. Diagnostic ability, therefore, results from a broad range of both academic and field experience mixed together with a great deal of common sense. The key ingredients of a good diagnostician are an inquisitive mind, a reluctance to jump to hasty conclusions, a persistent determination to find the cause of a problem, and a willingness to seek additional information from other tree specialists and reference texts.

A thorough understanding of the appearance and function of a healthy tree is necessary before a diseased tree can be recognized. The appearance throughout the year of each common tree species in an area should be known. A collection of field guides to woody ornamental plants is a helpful aid for rapid field identification. No one book is perfect for identification.

Knowledge of local conditions of weather and soil type is also necessary for accurate diagnosis of tree disease. The extremes of temperature and normal range of moisture conditions are important factors in both infectious and noninfectious disease. Knowledge of both recent history and past history of these conditions is needed, since trees often continue to respond to adverse environmental stress after the stress has stopped. The physical nature and chemical composition of soils are important because they will determine which tree species will adapt most successfully. Sometimes they are even the source of the trouble.

Diagnostic activities can be divided into field diagnosis and diagnosis by telephone, letter, or office visit. In field diagnosis the tree specialist examines the tree(s) and is usually able to question the client during the examination. This form of diagnosis allows the specialist to obtain the maximum information about the tree and site, from both direct examination and background information from the client. Samples for clinical studies in the laboratory may also be taken conveniently at this time. Diagnoses by telephone, letter, or office visit are more difficult because most or all the information comes from the client's impression of the problems. Samples of the tree sent or brought by clients are often not representative of the problem on the tree. In such cases, a tree's most serious problem may never come to light. Each of these forms of diagnosis, therefore, requires different approaches and they will be covered separately.

DIAGNOSIS OF TREE DISEASE IN THE FIELD

Introduction

Tree specialists must always remember that they are guests on the owner's property. Introduce yourself and let the owner point out the problem and escort you to the tree(s). Before you begin any diagnostic activity, notify the owner of your intentions. If samples of leaves, branches, bark, roots, or soil are needed, ask the client's permission before you begin sampling. Explain the reasons why a sample is needed and what disorder you suspect. Take as few samples as possible and try to avoid changing the appearance of the tree. Many owners are quite sensitive about their trees and do not want even a hopeless case mutilated by an overly aggressive diagnostician. Clean cutting tools with alcohol before and after use. Clean up after yourself when you are finished. Remove debris, replace soil or sod removed from around the tree, offer to apply wound dressing on wounds caused by sampling, and try to leave the area as you found it. Although none of the above is critical to accurate disease diagnosis, it is an expression of concern by the diagnostician for the client's trees and hence of concern for the client. This evidence of concern will foster confidence by the client in the diagnostician, which is essential if the client is to be expected to follow any recommendations.

Tree Examination

Tree examination should be a study of the total tree, including both below-ground and aboveground investigations, plus studies of the conditions around the tree and of the tree's history. The examination can begin at any point on or around the tree. Firm conclusions as to the cause of the problem should be deferred at least until the examination is finished.

Leaves, Branches, and the Trunk   The aboveground parts of the tree should be examined for the presence of necrotic tissues, abnormal growth of tissues, and evidence of pathogen structures. Necrosis is most easily detected by discoloration of plant tissues. Leaves become yellow or brown, inner bark on twigs and small branches changes from green to brown, and wood at the cambium turns from white to brown. Dead bark may also appear sunken as callus forms around the dead tissue. Pathogens frequently produce characteristic fruiting structures on recently dead tissue. Abnormal growths such as witches' brooms, galls, or any large swellings on a tree are often caused by pathogens. Some pathogens, such as the wilt fungi, are inside the wood and samples of branches must be taken to investigate for their presence. Whenever samples are taken from a tree, cutting tools should be surface-sterilized before the initial cut and before any new trees are sampled.

A key point in diagnosis is to be able to differentiate between organisms that kill plant tissues and then grow there, and organisms that merely grow on dead tissue. In most cases the pathogen is rapidly outcompeted and replaced by more aggressive saprophytes once the tissue is dead. Likewise, once a tree has been killed it may soon become difficult or even impossible to determine, or at least prove, what pathogen killed it. The best place to look for pathogens in a pure state is in the most recently killed tissues. Once tissues are dead, with cracked or peeled bark, the chances of observing or isolating the pathogen are greatly diminished. Even if the pathogen is still alive there, it will be mixed with other organisms. This distinction is obviously most critical when any samples for laboratory tests are taken. Most accurate diagnosis, therefore, can be achieved when a disease is in its earlier stages, before secondary pathogens and insects invade the host.

Roots   Root diseases are common in shade trees and the cause of disease symptoms on the leaves, branches, and the trunk can often be found in the condition of the roots. Indeed, such a cause may be entirely absent from the aboveground parts. The root system, however, is more difficult to examine than the aboveground parts of the tree. Sod, if any, should be removed carefully from around approximately half of the tree to about 2 ft (60 cm) from the trunk. Soil from around the roots should be removed carefully with hand tools. A narrow trowel and a whisk broom are most helpful. The condition of the roots should be checked by removing a small section of bark with a small knife. Wood beneath the bark of living roots will be creamy white while dead root tissues will have brown wood beneath the bark. The roots should also be checked for pathogen structures both on the root and beneath the bark. Decayed roots can also appear white but their soft, spongy texture makes them easy to detect. Soil should be removed until the spread of the root system from the trunk can be seen. The relationship of the root spread and the soil line should be determined to see if the tree is at the proper depth or is "deep planted." The presence and extent of girdling roots or cut roots can also be determined this way. This examination procedure is then repeated around the entire circumference of the tree.

Local Environment of the Tree   The location of the tree, the soil, and geographic features of the site can all give clues to stresses. Trees adjacent to roads, parking lots, sidewalks, driveways, and buildings, for example, are likely to be affected by people (people-pressure diseases). The physical characteristics of the soil around a tree may prevent adequate root growth. Shallowness or absence of topsoil, compacted soil, clay hardpans, and shallow rock ledges can be detected by soil examination. Measurement of moisture content of the top 18 inches (45 cm) of soil will reveal any possible moisture stress. Procedures for examining soil, and for taking samples for soil analysis to determine chemical composition, are given in Chapter 17, Soil Stress. A survey of local geography can indicate features that may contribute to environmental stress, such as frost pockets, areas of poor drainage, and severe wind exposure.

Information from the Client

After a thorough examination of a tree, including information from laboratory tests, the diagnostician often still remains uncertain. Additional information about local events and activities, even several years before, is needed for accurate diagnosis. This essential knowledge cannot be obtained through tree examination but must come from people familiar with the history of the tree(s). The person most likely to provide this information is the homeowner or grounds-keeper, unless the property recently changed hands.

There is no standard method of obtaining the needed information from a client. Have the client start talking about the tree and direct the conversation to possible causes of the problem. It is most useful to conduct questioning while the tree is being examined. Your questions will often help the client to remember past events that were forgotten. The client, however, will not usually associate certain important events, such as installation of a sewer line, driveway widen-ing, accidental chemical spills, and application of deicing salts, with tree disease. Questions about various types of activities that can be detrimental to trees, therefore, should be asked. Don't accept an "I don't think so" answer if reason-able evidence points to a particular activity around the tree. Politely urge the client to think harder, and suggest the possible occurrence of an activity similar to that suggested. For example, most clients will be aware of the dangers of brush killers around trees but are not aware that herbicides are also included in weed and feed mixtures.

Sometimes information may be withheld by the client. One explanation could be that an embarrassing mistake in tree care was made around the tree and the perpetrator hopes the diagnostician will blame the problem on some other cause. Sometimes withholding information is a way of "testing" the diagnosti-cian's skill. In these cases a bit of "detective" work is sometimes necessary to extract the information. Say "It looks like other cases of _____" and check the client's reactions. Most people will eventually reveal all the information if the diagnostician is probing, persistent, and polite.

The degree of involvement with the tree of the people questioned should also be determined. Beware of supervisors who are not acquainted with the problem firsthand; it is best to talk directly with those actually administering care to the trees. Beware also of secondhand information; much information becomes lost or confused as a message is passed along. Be careful to have authority from the actual owner to examine the tree. People who want their neighbor's tree examined may lead you into the cross-fire of a neighborhood feud.

DIAGNOSIS BY OFFICE VISIT, TELEPHONE, OR LETTER

There are many cases when diagnosis cannot be made at the tree. Most diagnosticians, especially those employed by federal and state governments, have large geographic areas to cover and cannot visit many trees. Requests for diagnosis and information are received and answered, therefore, by telephone, office visit, and letter. However, specimens brought or sent may not represent the principal affliction of the tree. The diagnostician, in these cases, works under a disadvantage when compared to field diagnosis, being obligated to depend completely on information (and sometimes samples) provided by the client. The diagnostician should mention the tentative nature of such distant diagnosis.

Office Visit

Clients visiting the diagnostician should be encouraged to talk freely about their tree and its problem, much as in the field. Being a good listener is the key to office and telephone diagnosis. Remain suspicious, however, of all information provided by the client. The species of tree affected is often the most important information needed, but clients sometimes do not know that they have given the wrong tree identification. The client's degree of expertise and the extent of observations therefore should be determined. For example, if the client describes a severe leaf problem, ask if the twigs and branches, trunk and roots, were also examined and if any other tree in the area had the same symptoms. Root injuries cause leaf problems. Sometimes a sample from the affected tree is brought to the office. These specimens are often helpful, especially to confirm the species affected, but, since most clients are inexperienced about trees, they will take a good sample for diagnosis only by chance.

Questioning may begin with helping to determine the tree species and gradually elicit information to make a *tentative* diagnosis. It is unwise to make any specific recommendations, unless a good sample was brought in. The client should be told what observations and/or what samples are needed, and how to collect them. Diagnostic questionnaires (see Table 25.1) are aids that help the client to gather the information needed to allow the diagnostician to make the most accurate diagnosis of the problem.

Telephone

Diagnosis by telephone is even more difficult than that by an office visit, since the client's facial expressions in response to questions cannot be studied and

TABLE 25.1

Diagnosis of Tree Health Problems

Accurate diagnosis of tree health problems requires much information about the affected tree(s) and the nearby environment. The following items of information will assist the diagnostician in an examination of tree health problems.

1. Provide the following background data on the affected tree(s): A. Species   B. Age (years)   C. Height   D. Trunk diameter [inches across at four feet (cm across at 1.4 m)]   E. Length of time in present site (years).

2. Describe the symptoms of the tree's health problem. Include details of any mushrooms, conks, molds, ooze (flux), or insects on the tree. Also describe any unusual appearance of foliage, twigs, flowers, fruits, seeds, branches, trunk, or roots.

3. Are there more trees of the same kind nearby? How near? Are these or other trees similarly affected?

4. Has the trouble appeared in previous years? If so, for how long?

5. Has this tree or have nearby trees been sprayed for insect or disease control? If so, when? What materials were used?

6. Have herbicides (weedkillers) been used in the vicinity? If so, how near? When? What materials were used?

7. Is there any evidence on trunk, roots, or branches of mechanical injuries from lawnmowers, automobiles, tools, or machinery? Has the tree received special treatment for this or other troubles? If so, when? By whom was the treatment administered?

8. Describe any unusual local weather conditions (hot or cold, wet or dry, etc.) in the present season or in previous seasons.

9. Is th tree shaded by buildings or other objects for part or all of the day? Is the tree in an exposed, windy location?

10. Describe the topsoil [upper 6 to 8 inches (15 to 20 cm)] around the tree (sandy, heavy clay, or intermediate). Is the soil wet or dry? Is it well drained or poorly drained?

11. Has fertilizer or other material been added to the soil around or above the roots? If so, when? What materials were applied?

12. Is there healthy turf or sod growing over the tree's roots?

13. Have any construction activities taken place near the tree in the past 10 years? How near? When? Describe the activities in detail. Does cement, asphalt, or other type of pavement occur near the tree? If so, what type? How near? How long has it been there? Has the soil near the tree been compacted by automobiles or other traffic?

14. Are gas, water, steam, sewer, or other pipes or conduits present in the ground under or near the tree? If so, have tests for leakage been made?

15. Is the normal outward flaring (buttress swell) of the roots visible at the base of the trunk? Does the tree trunk appear to go straight down into the ground like a telephone pole? Has the soil level around the tree been raised or lowered by fill or by grading operations during the past 10 years? If so, when? What was the change of level?

16. Check the base of the tree for girdling roots.

17. If some of the branches are wilted, examine the outer wood (sapwood) of the affected twigs for discoloration caused by vascular wilt pathogens. A laboratory culture of a suspect twig may be necessary to confirm this diagnosis and to help identify the pathogen.

---

specimens are not available. When people telephone, however, they often expect quick answers and are not aware that inaccurate diagnosis is likely. The diagnostician, therefore, must be careful to keep telephone diagnosis tentative. The client should be asked to complete and return a questionnaire that will give instructions on how to furnish samples and information needed for a more accurate diagnosis.

Letters

Letters are usually the most common source of request for diagnosis but in most cases provide the least amount of information. Specimens enclosed with the letters often arrive either dried out and broken or covered with surface mold. Even a tentative diagnosis is often difficult from information provided in many clients' letters. Again, questionnaires are most useful in getting the information needed.

TOOLS USED IN DISEASE DIAGNOSIS

A large number of tools have been used by diagnosticians to inspect the trees or soil and to collect samples for further testing. Some that the author has found useful in field diagnosis are listed in Table 25.2. Each diagnostician, however, develops a collection of tools that he or she finds most useful. Those with laboratory facilities will usually have a low power dissecting microscope and a high power microscope for closer examination of specimens collected.

Better tools for early disease diagnosis are needed. Control of many plant diseases often depends on early diagnosis. Recently, the plant pathologist has begun to use electronic technology to obtain additional objective information about the health of shade trees, just as the medical doctor has used such technology to detect and diagnose human diseases. Tattar and Blanchard (1977) discuss the use of nondestructive electrical measurements to detect and diagnose cold temperature injury, internal discoloration and decay, root rot, vascular wilt disease, air pollution injury, cankers, and mycoplasma disease in trees. These electrical measurements will not replace diagnostic procedures in current

TABLE 25.2
Tools Used in Disease Diagnosis

Camera	Alcohol in plastic bottle
Pole saw	Plastic bags [1 quart (liter) size or larger]
Pole pruner	Wound-dressing paint
Hand clippers	Trowel
Hatchet	Soil-sampling tube
Leather mallet	Soil auger
Pruning knife (curved, folding)	Gas detection meter
Increment borer	Shigometer and battery powered drill
Measuring tape	Calling cards
Diameter tape	Whisk broom
Hand lens	Cardboard tags with strings

use but will provide additional information to achieve more accurate detection and diagnosis. The Shigometer (Osmose Wood Preserving Co., Buffalo, NY), a field ohmmeter, is the first such electronic instrument designed for disease diagnosis of trees. This instrument is being used to detect diseases of shade trees in the field, such as internal discoloration and decay, and cankers. Additional electronic instruments will undoubtedly be developed in the future, permitting more accurate and objective diagnosis of disease in shade trees.

CONCLUSIONS

It may appear that the diagnosis of tree disease is often inaccurate. This is partially true but the odds for accurate diagnosis increase with practice and knowledge of the many tree diseases. Although general diagnostic procedures have been outlined, no standard procedure exists and each new situation may require a new approach. Don't be afraid to say "I don't know" and to seek help.

LITERATURE CITED

Tattar, T. A., and R. O. Blanchard. (1977). Electrical techniques for disease diagnosis. *J. Arbori.* **3**, 21–24.

SUGGESTED REFERENCES

Carter, J. C. (1959). Tools for field diagnosis. *Arborist's News* **12**, 89–92.
Chater, C. S. (1974). Diagnosis of shade tree problems. *Trees Mag.* **33**, 7–9.
Davis, S. H. (1974). How accurate are you in diagnosing your tree problems. *Arborist's News* **39**, 53–60.
Himelick, E. B. (1974). Fundamental diagnostic procedures in arboriculture. *Proc. Midwest Chapter Int. Shade Tree Conf.* **29**, 21–43.
Neely, D. (1972). Hints on diagnosing of tree problems. *Proc. Int. Shade Tree Conf.* **4**, 33–37.
Shigo, A. L., and A. Shigo. (1974). Detection of discoloration and decay in living trees and utility poles. *U.S., For. Serv., Res. Pap. NE* **NE-294**, 1–11.
Shigo, A. L., and P. Berry. (1975). A new tool for detection of decay associated with Fomes annosus in Pinus resinosa. *Plant Dis. Rep.* **59**, 739–742.
Streets, R. B. (1975). Diagnosis of plant diseases. Univ. of Arizona Press, Tucson.
Tattar, T. A. (1976). Use of electrical resistance to detect Verticillium wilt in Norway and sugar maple. *Can. J. For. Res.* **6**, 499–503.
Welch, D. S. (1960). The diagnosis of shade tree troubles. Midsummer Suppl. New York State Arborists Assoc., Ithaca, New York.

26

Living Hazard Trees

INTRODUCTION

Most people know that a dead tree should be removed as soon as possible, but few people realize that many living trees can be a threat to life and property. A living hazard tree is any tree that is structurally weakened so that all or parts of it are likely to fall. Unfortunately, living hazard trees are often overlooked. People usually become aware of such trees only after some major limb, or the whole tree, has fallen.

A tree's structural support is most important during high winds, or when snow and ice have built up on branches. A tree needs flexibility to withstand these periods of stress, which put great physical strain on a tree's own bulk. Because most tree failures happen during a wind storm, or when snow and ice accumulate, people usually think of them as unpreventable accidents. Many times, however, chronic diseases or structural defects cause trees to fail during storm conditions. In these cases, the tree "accident" could have been prevented. This chapter examines the most common types of living hazard trees to encourage early detection and removal of such trees.

BACKGROUND

Dr. Lee Paine has conducted considerable research on living hazard trees in national forests and parks in the western United States. A concern for public safety prompted his research. He found that, all too often, both recreation area officials and visitors were unaware of the potentially dangerous trees around them. However, recreation area managers, municipal tree officials, and homeowners can be held liable for injuries and losses due to preventable hazard tree failures on property under their control (Fig. 26.1). How does a sound, healthy, living tree become a hazardous one? A cross-section of the trunk of a healthy tree shows that nearly all of the volume of the tree is wood (xylem) (Fig. 26.2).

Fig. 26.1 Tree failure from internal defects that has resulted in property loss. The tree owner could be liable for this damage.

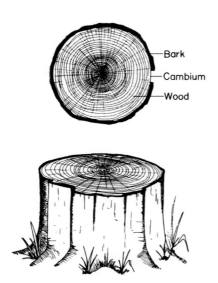

Fig. 26.2 Cross section of a living tree.

Most of this wood is made of dead supporting tissues; only the tissues in the narrow band nearest the cambium are alive. If the woody tissues inside the cambium become unsound for some reason, such as internal decay, then much or most of the structural support for the tree will be lost. The tree, however, will remain alive, because the living cells near the cambium will continue to grow so the tree can function normally. In fact, these living cambial cells continue to lay down new xylem layers. The result is a normal- and healthy-looking tree that nonetheless does not have enough structural strength (from wood) to support itself during periods of stress, such as wind or ice storms.

CATEGORIES OF HAZARDOUS TREES

Living trees become structurally unsound in four common ways: (1) internal decay in the trunk and/or large branches; (2) cankers and canker-rots; (3) cut roots and root decay; and (4) weak forks in the trunk and/or large branches.

Internal Decay

Decay in living trees is the end result of many complex interactions between the tree and several groups of microorganisms (principally fungi and bacteria). All these processes begin with a wound, through which microorganisms enter. Most trees suffer broken branches, and many undergo other mechanical injuries that leave wounds for decay-producing microorganisms to enter. Most healthy and vigorous trees have adequate systems of protection and repair, which limit this invasion, so these trees remain structurally sound. However, when trees are in poor health and low vigor, the advantage goes to the microorganisms and extensive columns of decay often result (Fig. 26.3). Once decayed tissue takes over a large volume of the trunk or major branches, which were formerly filled by sound wood, the tree will be unable to support even its own weight (Fig. 26.4). Although most tree failures occur during storms, occasionally such a tree will fail on a completely calm day.

The advancing columns of discoloration followed by decay do not spread to the cambium or bark and, therefore, do not kill the tree. The tree is able to wall off or "compartmentalize" the invading microorganisms in woody tissues that had already been formed before wounding (see Chapter 13, Wound Diseases). Unfortunately, a tree in poor health grows very slowly, and its systems of protection and repair cannot prevent a large amount of decay from occurring within a relatively short period. In addition, such a tree is likely to have a few branches die each year, which creates new wounds and allows new zones of decay to get started. The result is a tree with a narrow band of living and functional wood that keeps the branches and leaves alive, but contains extensive decay that may cause the tree to fail *at any time* (even though failure is most likely to occur during high winds).

Fig. 26.3 Sugar maples in poor vigor that have considerable internal defects. Note dead branches in the crown of both trees, and fruiting bodies of decay fungi (arrow) on a branch of tree (B). Note also extensive bark killing from wounds in tree (A).

Fig. 26.4 Failure of a sugar maple from extensive internal defects. (Photo courtesy of Shade Tree Laboratories, University of Massachusetts, Amherst.)

Fig. 26.5 Vertical bark cracks on sweetgum. Note excessive callus formed along crack from many years of attempted closure by the tree.

Trees that are potentially hazardous as a result of internal decay often exhibit external evidences of their internal weakness that are helpful in diagnosis. Large dead and dying branches throughout the crown often indicate a tree in low vigor that may be hazardous for a number of reasons (Fig. 26.3). Large and deep vertical cracks on the trunk or large branches are often early indications that failure from internal decay is possible (Fig. 26.5). If cracks are located at opposite sides of the tree, failure may be imminent. Large areas of exposed wood on the trunk, even if they appear sound, indicate serious older wounds that have not closed and where the potential for extensive internal decay exists (Fig. 26.6). Large columns of decay can often be found beneath this hard exterior. Old branch wounds that remain open usually indicate serious decay inside the tree (Fig. 26.7). The presence of conks, mushrooms, or bracket fungi on the trunk are all evidence that decay fungi are at work inside the tree (Fig. 26.8). Carpenter ants do not attack sound wood, but follow existing columns of decayed wood. Their presence in large numbers around a wound, therefore, is evidence of internal decay. If any of these indicators are found during routine inspection of a tree, a thorough hazard tree inspection is warranted.

Fig. 26.6 Neil Calvanese of the Central Park Conservancy examines the extensive decay that has followed a severe bark wound on an American sycamore.

Fig. 26.7 Old unclosed wounds on (A) horsechestnut and (B) American elm. Note slime flux below wound on elm.

Cankers

Cankers, which are localized dead areas on the bark (see Chapter 9, Stem Diseases), can also lead to structural instability. Cankers develop when micro-organisms kill the cambium so the tree cannot close any wound it might suffer from broken branches or mechanical injuries. Canker pathogens, usually fungi, are also able to invade adjacent healthy bark and establish an enlarging peren-nial infection. Since the tree grows only at the living cambium layer, the dead, canker-infected tissue begins to appear sunken into the stem or branch as annual ridges of callus form around it (Fig. 26.9).

A healthy tree can bend and sway with the wind without breaking because the wood fibers slide past each other. This flexibility of the stem and branches is most important in the "holding wood" [outer 3 to 6 inches (7 to 15 cm) of wood] of the tree. Trees in which cankers have killed large areas of bark do not have

Fig. 26.8 Conks of decay fungi serve as indicators of internal decay on these two black locusts.

much flexibility. Trees with large cankers often break at the canker face and should thus be removed (Fig. 26.10).

Canker-Rots

Some microorganisms attack and decompose the wood beneath the canker as well as the neighboring bark tissues. If the decay progresses rapidly, the resulting condition is termed a *canker-rot* (see Chapter 9). Often this decay is on the canker face and easy to detect. But, in some cases, it is mostly internal and not visible from the outside of the tree. Most canker-rots are likely to cause tree failure because of the combined effects of (1) dead bark around the circumference of the tree (from the canker) and (2) loss of internal support (from wood

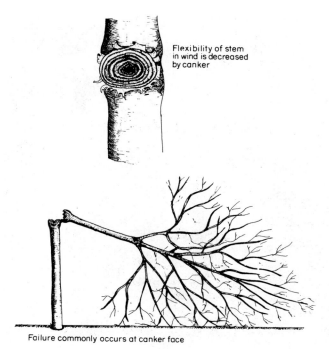

Flexibility of stem
in wind is decreased
by canker

Failure commonly occurs at canker face

Fig. 26.9   Tree failure from a stem canker.

decay). A canker-rot is a very hazardous condition, and affected trees should be removed as soon as possible.

Root Failure

Roots hold a tree firmly in place. Anything that alters the structural support provided by any part of the root system decreases the stability of the tree. Root decay (root rots) (see Chapter 10, Root Diseases) and severing or cutting any portion of the root system are two factors that decrease a root system's ability to support the tree. In addition, soil erosion, drought, gas leaks, fill, flooding, soil compaction, or paving over roots can kill roots.

Roots decay when attacked by microorganisms, which are usually found in the soil or growing on the roots of a nearby tree or stump. Microorganisms often infect a root through a wound, but some are able to penetrate an intact root directly, especially after a tree has been weakened, for example, by insect defoliation. Some root rot microorganisms kill certain tree species before the tree has been weakened enough to fall. However, many root rots cause living trees to fall. Entire trees are often blown over with the remains of their decayed root systems still attached (Fig. 26.11).

Fig. 26.10   Severe Eutypella trunk canker on sugar maple.

Trees infected with root rot fungi often have visible fruiting structures of the fungus (mushrooms or conks) on the lower trunk (Fig. 26.12). When extensive infections are detected, the tree should be removed. Occasionally the micro-organisms that cause roots to decay also invade the root buttress and the lower stem (bole) of the tree. In these cases, the tree may fail at those areas as well as at the roots.

If a fallen tree has an abnormally small root system with several decayed roots, there can be little doubt that root rot caused the tree failure. Sometimes unin-fected trees blow down because they have a shallow root system. Trees that grow on rock ledges, near a body of water, or on ground with a high water table often have shallow root systems and can be living hazard trees. In addition,

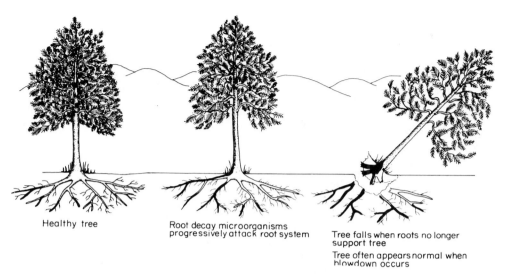

Healthy tree

Root decay microorganisms
progressively attack root system

Tree falls when roots no longer
support tree

Tree often appears normal when
blowdown occurs

Fig. 26.11 Tree failure from root rot.

Fig. 26.12 Fruiting bodies of decay fungi on the roots of (A) sugar maple and (B) oak that serve as indicators of root rot.

Fig. 26.13   Cut roots on a red oak due to earthmoving equipment.

trees that formerly grew in a dense stand of trees, most of which have been cut for road or building construction, are also in high danger of root failure.

Building, road, and sidewalk construction and pipeline installation are especially hazardous for the roots of nearby trees. Earth-moving and trenching equipment used around trees often severs a large portion of the tree's roots (Fig. 26.13). Roots may also be cut during excavation for a cellar hole or for septic tank installation. Without the support of its entire root system, the tree is structurally weakened. The probability of failure increases with the amount of the root system cut. Trees that have lost 50% or more of their root system during construction should be removed.

Wounds created by construction equipment are often invaded by the root rot microorganisms discussed previously. The interaction of root injury and root rot microorganisms can cause additional tree failure years after buildings, roads, pipelines, or sidewalks have been completed.

Weak Branches or Forks

As trees in a forest grow and compete for sunlight, they usually lose their lower branches early in life. Forest trees normally grow straight and tall with

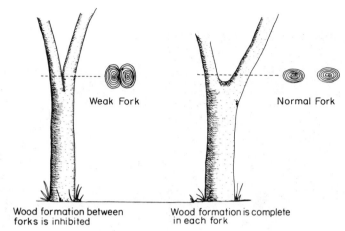

Weak Fork   Normal Fork

Wood formation between   Wood formation is complete
forks is inhibited   in each fork

Fig. 26.14 Normal and weak forks.

narrow crowns. Shade trees, however, which are grown in the open along streets, in parks, and around homes, are not usually forced by competition for sunlight to grow rapidly upward. These trees keep their lower branches and develop wide and flat crowns that yield large areas of shade beneath them. One disadvantage for the tree of this wide crown-growth habit may be the development of large, weak branches. When a branch or fork forms, depending upon tree species or variety, there is likely to be a 45° to 90° angle between the trunk and the branch. In this case enough wood forms around both the trunk and branch to support the weight of both. However, if this angle is much narrower (less than 40°), not much supporting wood will form on the inside of the angle because of the pressure exerted during growth from both sides of the fork (Fig. 26.14). This makes the fork structurally weak. As the stem's weight continues to increase, the weak fork will usually split at this junction, often resulting in the failure of a branch or even a large portion of the crown (Fig. 26.15).

Some tree species, such as chestnut, horse chestnut, maple (especially silver maple), mimosa, oak, poplar, sassafras, Siberian elm, tuliptree, and willow, are often susceptible to weak forks. These species also may be more likely to break because of their inherited branch-growth habits or because their wood is brittle. These trees should be examined closely for potentially weak forks, especially when they are growing in areas where failure could be most damaging. Weak branches should be supported by installing cables or braces, or should be removed.

A weak fork may not split completely at first, but may only open a fissure (crack) that can be invaded by microorganisms (Fig. 26.16). The resulting decay further weakens the fork and hastens its failure.

Fig. 26.15 (A) Failure of one or two main branches of a sugar maple from a weak fork; (B) close-up. Note remaining main branch at right and lack of internal defects in the wood.

DETECTION OF LIVING HAZARD TREES

Most failures of shade trees could be prevented if efforts were focused on early detection. However, since such an effort cannot be extended to all trees, the available resources should be concentrated on trees in intensively used areas along roadsides, in recreation sites, and around homes. Arborists must be able to recognize the symptoms of the most common examples of internal defects, decays, cankers, canker-rots, and root rots to detect living hazard trees. In addition, any building, road, pipeline, and/or sidewalk construction near trees means that trees could suffer cut or injured roots. These trees should be watched for the next 10 to 20 years for any signs of weakness. Carefully examine trees with potentially weak forks for signs of cracking or internal decay. In cases where cabling or bracing are not possible, it may be necessary to remove the weak fork or even the entire tree.

There are times when the extent of a defect cannot be seen from the outside. In cases where there is likely to be decay in the trunk or large branches, more information about the exact amount of structural weakness is needed before a decision about removal or preservation can be made. Two tools are commonly used as aids for detecting internal decay in living trees: (1) the increment borer and (2) the Shigometer.

Fig. 26.16 Crack formed at weak fork on American elm. Note bleaching of bark below the crack from slime flux.

The increment borer cuts a small cylinder of wood from the tree. This cylinder is examined visually or with the aid of a hand lens or microscope to determine how much sound (nondecayed) wood remains in the tree. In addition, a recent history of growth increments of the tree can obtained from the increment core(s). Unfortunately, the increment boring procedure itself produces wounds that can eventually lead to extensive discoloration and decay, and its repeated use on the same tree could be quite damaging. However, the long-term wound effects of any type of investigative borings must be balanced against the concern for human safety and property damage. Responsible arborists will make test borings only when necessary, to enable them to make the most accurate recommendation concerning the fate of a suspected hazard tree.

The Shigometer uses the resistance to a pulsed electric current to measure the extent of discoloration and decay inside a tree. A small drill hole (much smaller than the increment borer hole) is made in the tree, and an electrode probe is inserted (Fig. 26.17). The presence and severity of discoloration and decay can be determined by changes in electrical resistance measurements. In general, clear tissue has a high resistance, while discolored wood and decayed wood have progressively much lower electrical resistances. Resistance in an air-filled cavity, on the other hand, is infinitely high, since there is no wood present and

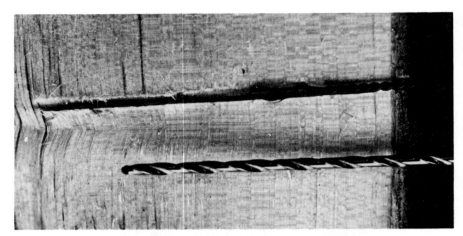

Fig. 26.17 Hole in wood made with $\frac{3}{32}$-inch drill (above). Note twisted-wire electrode in hole that is used to determine electrical resistance of wood in living trees with a Shigometer. (Photo courtesy of Alex L. Shigo, USDA Forest Service, Durham, New Hampshire.)

the instrument measures only air. In the hands of a trained professional the Shigometer is an accurate diagnostic tool for detecting internal decay in trees. It is unfortunate that the high price of the instrument has limited its use by arborists.

CONCLUSIONS

Many tree failures are preventable. Most are the result of chronic disease and/or structural defect problems. Since many failures are predictable as well as preventable, property owners and municipalities might be considered negligent for damage caused by some tree failures. In addition, losses from such damage may not quality as a federal income tax deduction because this covers only sudden, unexpected, or unusual property losses. Property owners and municipalities should take reasonable steps to detect and remove potentially hazardous trees. Periodic examination of such trees should be performed by tree professionals.

SUGGESTED REFERENCES

Blanchard, R. O., and T. A. Tattar. (1981). Field and laboratory guide to tree pathology. Academic Press, New York.
Paine, L. A. (1971). Accident hazard evaluation and control decisions on forested recreation sites. *U.S. For. Serv., Res. Paper PSW* **69**, 1–10.

Paine, L. A. (1973). Administrative goals and safety standards for hazard control on forested recreation sites. *U.S. For. Serv., Res. Paper PSW* **88,** 1–13.

Shigo, A. L., and H. G. Marx. (1977). Compartmentalization of decay in trees. *USDA For. Serv. Agric. Info. Bull.* **405,** 1–73.

Shortle, W. A. (1979). Detection of decay in trees. *J. Arbori.* **5,** 226–232.

Glossary

Aerobic. Growth in the presence of oxygen

Anaerobic. Growth in the absence of oxygen

Antibiotic. A chemical produced by a microorganism that is inhibitory or toxic to other microorganisms

Asexual. Reproduction without the union of male and female reproductive cells

Bacterium. A single-celled microorganism with a cell wall that lacks chlorophyll and multiplies by fission

Bleeding. Sap flow from a wound

Blight. A rapid killing of large areas of the tree

Blotch. A large irregular necrotic area on the leaf that diffuses into healthy tissues

Cambium. A one- or two-celled layer between the phloem and xylem that produces both these tissues and results in diameter growth

Canker. A localized necrotic lesion on the bark

Chlorosis. Loss of green color of leaves due to the inability to form chlorophyll

Conk. A sporophore of a wood decay fungus

Cultivar. A selected and propagated variety of a plant

Dieback. Progressive death of branches beginning at the terminals usually starting in the upper crown

Disease. Any condition that adversely affects the health of a plant

Disease cycle. The sequence of events involved in the development of a pathogen during the progression of an infectious disease

Eradication. Control of disease by removal of infected parts or by removing infected individuals in a population

Exclusion. Control of disease by preventing infected plants from coming into an area free of the disease

Facultative parasite. A saprophytic organism that can become parasitic under certain conditions

Fertilization. Combining of two gametes resulting in a doubling in chromosome number

Flag. A single dying branch in a healthy crown

Fleck. A minute spot

Fruiting body. A structure composed of mycelium that contains spores

Fumigant. A gas or volatile compound used to sterilize soil

Fungicide. A compound that kills fungi

Fungistat. A compound that stops a fungus from growing but does not kill it

Fungus A nongreen plant with a vegetative body composed of hyphae that reproduces by spores

Gall. A pathogen-induced swelling or overgrowth

Giant cell. A large multinucleate cell formed from the combining of several adjacent cells usually in response to infection by root knot nematodes

Haustorium. A specialized structure capable of direct penetration into the plant and subsequent absorption

Honeydew. Sweet, sticky excretion of aphids and some other sucking insects

Host. A plant that provides nutrition for an invading parasite

Hydathodes. Structures with openings at the end of a vein in a leaf that discharge water

Hypha. A single hollow tube of fungus growth

Infection. An establishment of a food relationship between a host and a parasite

Infectious disease. A disease caused by a pathogen capable of eventual spread to another plant

Inoculate. Bringing the pathogen into contact with a susceptible plant

Invasion. Spread of the pathogen within the host

Leaf spot. A well-defined necrotic area on a leaf

Lenticel. A structure on the bark that allows gas exchange

Lesion. Area of dead tissue

Local lesion. A spot caused by the mechanical inoculation of a virus

Mildew. A fungus mycelium and spores on the surface of the plant usually appearing white or gray

Mold. A profuse growth of fungus mycelium on dead plant tissue or any dead matter

Mosaic. Alternate patches of normal and light-green or yellow on leaves

Mottle. An irregular pattern of light and dark green areas on leaves

Mushroom. A fruiting body of many wood decay fungi that contains gills on the underside

Mycelium. A mass of fungus hyphae

Mycorrhiza. A symbiotic association between a fungus and a feeder root of a plant

Natural openings. Hydathodes, lenticels, and stomata

Necrotic. Dead

Nematocide. A compound that is toxic or inhibitory to nematodes

Nematode. A roundworm with a cuticle and a hydrostatic skeleton that is often a parasite on plants or animals

Noninfectious disease. A disease caused by stress from the environment, macroscopic animals, or people, which cannot be passed from a diseased plant to a healthy plant

Nucleic acid. Compounds found in the chromosomes and other cell constituents that control gene expression, DNA (deoxyribonucleic acid), and RNA (ribonucleic acid)

Obligate parasite. A parasite that can only grow and reproduce on living organisms

Parasite. An organism that lives on or in another organism (host) and obtains its food from that organism

Pathogen. An agent that causes disease

pH. Degree of acidity or alkalinity, expressed as the reciprocal logarithm of the amount of hydrogen ions

Phloem. Food-conducting tissue immediately outside the cambium, known collectively as the inner bark

Phytotoxic. Toxic to plants

Predisposition. Influence of numerous living and nonliving stress factors on the susceptibility of a plant to disease

Protection. Application of materials that prevent infection by pathogens

Pustule. A small blister-like swelling caused by a fruiting body beneath the plant surface

Resinosis. Exudation of pitch from a wound or infection on a conifer

Resistance. Ability to retard the development of a disease

Rot. Decay or physical decomposition of tissues

Sanitation. Removal and destruction of infected parts or individuals; decontamination of tools, clothing, or working surfaces

Saprophyte. An organism that feeds on dead organic matter

Scab. A dark, crustlike spot on a leaf or fruit

Scorch. Necrosis of leaf margins or tips

Sexual. Reproduction involving the union of male and female reproductive cells

Shot-hole. A symptom produced by the dropping out of the necrotic center of a leaf spot leaving a hole in the affected leaf

Species. A group of individuals that have characteristics in common and are closely related genetically

Spore. The reproductive unit of the fungi, which may be one to many cells

Stomate. A minute opening in the leaf surface that allows gaseous exchange

Strain. A relatively pure genetic line of organisms

Stroma. A mass of hyphae that contain fruiting bodies

Stylet. A hollow feeding tube of a sucking insect or a plant parasitic nematode

Symptom. Visible response of a plant to a disease

Systemic. Movement through the vascular system, usually by a pathogen or chemical

Tolerance. Ability to withstand the adverse effects of an infectious or noninfectious disease with minimal injury

Variety. A subgroup within a species

Vascular. Conducting tissues, phloem and xylem

Vector. An organism able to transmit a pathogen

Vessel. A water and mineral conducting element in the xylem

Virulence. A measure of ability to cause disease of a pathogen

Virus. A submicroscopic parasite composed of a protein coat and a nucleic acid core that can only reproduce in living tissue

Wilt. Drooping of foliage from lack of water

Witches' broom. The result of dense, prolific branching on a branch or stem, usually associated with a pathogen

Xylem. Water and mineral conducting tissue; wood

Appendix I

Common and Scientific Names of Ornamental Woody Plants

ailanthus, *Ailanthus altissuma*
alder, black, *Alnus glutinosa*
alder, Japanese, *A. japonica*
alder, red, *A. rubra*
alder, speckled, *A. incana*
almond, flowering, *Prunus triloba*
althea, shrubby, *Hibiscus syriacus*
amelanchier, *Amelanchier*
apple, *Malus*
apricot, *Prunus*
arborvitae, American, *Thuja occidentalis*
arborvitae, Japanese, *T. standishii*
arborvitae, giant, *T. plicata*
ash, Chinese, *Fraxinus chinesis*
ash, green, *F. pennsylvanica*
ash, white, *F. americana*
aspen, European, *Populus tremula*
aspen, largetooth, *P. grandidentata*
aspen, trembling, *P. tremuloides*
avocado, *Persea*
barberry, *Berberis*
basswood, American, *Tilia americana*
beech, American, *Fagus grandifolia*
beech, European, *F. sylvatica*
birch, European, *Betula pendula*
birch, gray, *B. populifolia*
birch, paper, *B. papyrifera*
birch, water, *B. occidentalis*
birch, yellow, *B. alleghaniensis*
blackberry, *Rubus*

boxelder, *Acer negundo*
boxwood, *Buxus*
buckeye, California, *Aesculus californica*
buckeye, Ohio, *A. glabra*
buckeye, red, *A. pavia*
buddleia, *Buddleia*
butternut, *Juglans cinerea*
camellia, *Camellia*
catalpa, *Catalpa*
cedar, atlantic white, *Chamaecyparis thyoides*
cedar, northern white, *Thuja occidentalis*
cedar, red, *Juniperus*
cherry, black, *Prunus serotina*
cherry, choke, *P. virginiana*
cherry, sour, *P. cerasus*
cherry, sweet, *P. avium*
chestnut, American, *Castenea dentata*
chestnut, Chinese, *C. mollissima*
chestnut, Japanese, *C. crenata*
chestnut, Spanish, *C. sativa*
clematis, *Clematis*
coconut palm, *Cocos nucifera*
cotoneaster, *Cotoneaster*
cottonwood, eastern, *Populus deltoides*
cottonwood, northern black, *P. trichocarpa*
crabapple, *Malus*
cryptomeria, *Cryptomeria japonica*
dawn redwood, *Metasequoia glyptostroboides*
deutzia, *Deutzia*
dogwood, flowering, *Cornus florida*

dogwood, red, *C. sanguinea*
dogwood, red-osier, *C. stolonifera*
Douglas fir, *Pseudotsuga menziesii*
elder, *Sambucus*
elm, American, *Ulmus americana*
elm, cedar, *U. crossifolia*
elm, Chinese, *U. parvifolia*
elm, English, *U. procera*
elm, red (slippery), *U. rubra*
elm, September, *U. serotina*
elm, Siberian, *U. pumila*
elm, winged, *U. alata*
eucalyptus, *Eucalyptus*
euonymus, *Euonymus*
exochorda, *Exochorda*
fig, *Ficus*
fir, balsam, *Abies balsamea*
fir, grand, *A. grandis*
fir, white, *A. concolor*
firethorn, *Pyracantha*
forsythia, *Forsythia*
ginkgo, *Ginkgo biloba*
Golden-rain tree, *Koelreuteria paniculata*
grape, *Vitis*
gum, black, *Nyssa sylvatica*
hackberry, *Celtis*
hawthorn, *Crataegus*
hemlock, eastern, *Tsuga canadensis*
hemlock, western, *T. heterophylla*
hickory, shagbark, *Carya ovata*
holly, *Ilex*
honeylocust, *Gleditsia triacanthos*
horsechestnut, *Aesculus hippocastanum*
Japanese pagoda-tree, *Sophora japonica*
jasmine, *Jasminum*
juniper, Chinese, *Juniperus chinensis*
juniper, common, *J. communis*
juniper, creeping, *J. horizontalis*
Katsura-tree, *Cercidiphyllum japonicum*
Kentucky coffee tree, *Gymnocladus dioicus*
Kerria, *Kerria japonica*
larch, American, *Larix laricina*
larch, European, *L. decidua*
larch, Japanese, *L. leptolepis*
larch, western, *L. occidentalis*
lilac, *Syringa*
linden, American, *Tilia americana*
linden, Chinese, *T. tuan*
linden, common European, *T. europaea*
locust, black, *Robinia pseudoacacia*
locust, honey, *Gleditsia triacanthos*
locust, Moraine, *G. triacanthos*

London plane tree, *Plantanus acerifolia*
magnolia, cucumber tree, *Magnolia acuminata*
magnolia, Dawson, *M. dawsoniana*
magnolia, lily-flowered, *M. liliiflora*
magnolia, southern, *M. grandiflora*
mahogany, *Melia*
maidenhair tree, *Ginkgo biloba*
maple, black, *Acer nigrum*
maple, boxelder, *A. negundo*
maple, Japanese, *A. palmatum* and *A. japonicum*
maple, Norway, *A. platnoides*
maple, red, *A. rubrum*
maple, silver, *A. saccharinum*
maple, sugar, *A. saccharum*
maple, sycamore, *A. pseudoplantanus*
mimosa, *Albizzia julibrissin*
mountain-ash, American, *Sorbus americana*
mountain-ash, European, *S. aucuparia*
mulberry, *Morus*
nandina, *Nandina domestica*
oak, black, *Quercus velutina*
oak, blackjack, *Q. marilandica*
oak, California black, *Q. Kelloggii*
oak, English, *Robur*
oak, Gambel, *Q. gambelii*
oak, live, *Q. virginiana*
oak, northern red, *Q. rubra*
oak, pin, *Q. palustris*
oak, post, *Q. stellata*
oak, scarlet, *Q. coccinea*
oak, turkey, *Q. cerris*
oak, water, *Q. nigra*
oak, white, *Q. alba*
oak, willow, *Q. phellos*
oleander, *Nerium oleander*
osage-orange, *Maclura pomifera*
pagoda tree, Japanese, *Sophora japonica*
paulownia, *Paulownia tomentosa*
peach, *Prunus*
pear, *Pyrus*
pecan, *Carya pecan*
pepper-tree, Monks', *Vitex agnus-castus*
persimmon, *Diospyros*
pine, Austrian, *Pinus nigra* var. *austriaca*
pine, Caribbean, *P. caribaea*
pine, Coulter, *P. coulteri*
pine, eastern white, *P. strobus*
pine, foxtail, *P. balfouriana*
pine, jack, *P. banksiana*
pine, Japanese black, *P. thunbergiana*

pine, Jeffrey, *P. jeffreyi*
pine, Knobcone, *P. attenuata*
pine, Lawson, *P. lawson*
pine, limber, *P. flexilis*
pine, loblolly, *P. taeda*
pine, lodgepole, *P. contorta*
pine, longleaf, *P. palustris*
pine, Macedonian, *P. peuce*
pine, Monterey, *P. radiata*
pine, mugo, *P. mugo*
pine, oriental white, *P. parviflora*
pine, piñon, *P. edulis*
pine, pitch, *P. rigida*
pine, ponderosa, *P. ponderosa*
pine, red, *P. resinosa*
pine, sand, *P. clausa*
pine, Scots, *P. sylvestris*
pine, short leaf, *P. elliottii*
pine, sugar, *P. lambertiana*
pine, Swiss stone, *P. cembra*
pine, Table-mountain, *P. pungens*
pine, Torrey, *P. torreyana*
pine, Virginia, *P. virginiana*
pine, western white, *P. monticola*
pine, whitebark, *P. albicaulis*
plane tree, *Platanus*
plum, *Prunus*
pomergranate, *Punica granatum*
poplar, balsam, *Populus balsamifera*
poplar, European black, *P. nigra*
poplar, white, *P. alba*
privet, *Ligustrum*
quince, *Cydonia oblonga*
quince, flowering, *Chaenomeles*
raspberry, *Rubus*
redbud, *Cercis*
redcedar eastern, *Juniperus virginiana*
redcedar, western, *Thuja plicata*

rhododendron, *Rhododendron*
rose, *Rosa*
rubber tree, *Hevea brasiliensis*
Russian olive, *Elaeagnus*
sassafras, *Sassafras albidum*
serviceberry, *Amelanchier*
shadbush, *Amelanchier*
silk tree, *Albizzia julibrissin*
soapberry, *Sapindus*
sourgum, *Nyssa*
sourwood, *Oxydendrum arboreum*
spirea, *Spirea*
spruce, black, *Picea mariana*
spruce, blue, *P. pungens*
spruce, Engelmann, *P. engelmannii*
spruce, Norway, *P. abies*
spruce, red, *P. rubens*
spruce, Sitka, *P. sitchensis*
spruce, white, *P. glauca*
sumac, *Rhus*
sweet gum, *Liquidambar styraciflua*
sycamore, American, *Plantanus occidentals*
tamarack, *Larix laricina*
thornapple, *Crataegus*
tree-of-heaven, *Ailanthus altissima*
tulip tree, *Liriodendron tulipifera*
tupelo, black, *Nyssa sylvatica*
umbrella-tree, *Melia azedarach*
walnut, black, *Juglans nigra*
walnut, English (Persian), *J. regia*
weigela, *Weigela*
willow, black, *Salix nigra*
willow, weeping, *S. babylonica*
yellow poplar, *Liriodendron tulipifera*
yellow-wood, *Cladastris lutea*
Yew, *Taxus*
zelkova, *Zelkova*

Appendix II

Use of Pesticides

The use of fungicides, herbicides, insecticides, miticides, nematicides, or any other biocides is now strictly regulated by both state and federal regulations. These laws require the potential applicator of pesticides to pass one or more examinations about the proper and safe use and disposal of pesticides. Before following any control recommendations involving the application of pesticides, look on the publication for its date and then check a current list of pesticides registered for that particular tree health problem. Pesticide registrations are often subject to frequent change; don't rely on your memory for either selection or dosage and don't rely on old literature. Always check the label before preparing a pesticide mixture.

Many pesticides currently registered for use on shade trees may be banned by the Environmental Protection Agency (EPA) in the near future. The source of the most current pesticide information is the state pesticide coordinator, usually located at the state university, or the nearest EPA office. Extension specialists in arboriculture, entomology, forestry, and plant pathology also try to keep up to date on the latest regulations. Annual Cooperative Extension Service publications on the control of diseases and pests of shade and ornamental woody plants are the best sources of written information on currently registered pesticides. Pesticide recommendations in textbooks or any older literature are often no longer valid. The following are three examples of extension publications that contain current recommendations for pesticide use of shade trees:

"Year Insect and Disease Control Guide for Trees and Shrubs." Cooperative Extension Service, University of Massachusetts. (Issued annually.)

"Year Recommendations for Chemical Control of Tree and Shrub Diseases and Nematodes in New York State." Cooperative Extension Service, NY. State College of Agriculture at Cornell University. (Issued annually.)

Cornell Recommendation for Commercial Production and Maintenance of Trees and Shrubs. NY State College of Agriculture. Cornell Miscellaneous Bulletin. (Issued in even numbered years.)

Appendix III

Literature Containing Information about Shade Tree Diseases or Insects

BOOKS

Agrios, G. N. (1988). Plant pathology, 3rd Ed. Academic Press, San Diego. 803p.

Baker, W. L. (1972). Eastern forest insects. USDA Forest Service Publ. #1175. 642p.

Blanchard, R. O., and T. A. Tattar. (1981). Field and laboratory guide to tree pathology. Academic Press, New York. 285p.

Boyce, J. S. (1961). Forest pathology. McGraw-Hill, New York, 572p.

Carter, J. C. (1975). Diseases of midwest trees. Illinois Natural History Survey, Urbana, Spec. Publ. 35. 168p.

Bega, R. V. (1978). Diseases of Pacific Coast conifers. USDA Agric. Handb. 521, 204p.

Benyus, J. M. (Ed.). (1983). Christmas tree pest manual. USDA Forest Service, North Central Forest Experiment Station. 108p.

Bernatzky, A. (1978). Tree ecology and preservation. Elsevier Science Publishing Co., New York. 357p.

Coyier, D. L., and M. K. Roane. (1986). Compendium of rhododendron and azalea diseases. American Phytopathological Society Press, St. Paul, MN. 65p.

Fosberg, J. L. (1975). Diseases of ornamental plants. University of Illinois College of Agric. Spec. Publ. No. 3 (Rev.) 220p.

Harris, R. W. (1983). Arboriculture-care of trees, shrubs, and vines in the landscape. Prentice-Hall, Englewood Cliffs, NJ. 688p.

Hepting, G. H. (1971). Diseases of forest and shade trees of the United States. USDA Forest Service Handbook #386. 658p.

Johnson, W. T., and H. H. Lyon. (1976). Insects that feed on trees and shrubs. Cornell University Press, Ithaca, NY. 464p.

Horst, R. K. (1979). Wescott's plant disease handbook, 4th Ed. Van Nostrand, New York. 803p.

Manion, P. D. (1981). Tree disease concepts. Prentice-Hall, Englewood Cliffs, NJ. 400p.

Partyka, R. E., J. W. Rimelspach, B. G. Joyner, and S. A. Carver. (1980). Woody ornamentals. Chemlawn Corp., Columbus, OH. 427p.

Peace, T. R. (1962). Pathology of trees and shrubs with special reference to Britain. Clarendon Press, Oxford.

Peterson, G. W., and R. S. Smith. (1975). Forest nursery diseases in the United States. USDA Forest Service Handbook No. 470. 125p.

Phillips, D. H., and D. A. Burdekin. (1982). Diseases of forest and ornamental trees. MacMillan, London.

Pirone, P. P. (1978). Tree maintenance, 5th Ed. Oxford University Press, New York. 587p.

Pirone, P. P. (1978). Diseases and pests of ornamental plants. 5th Ed. John Wiley & Sons, New York. 566p.

Powell, C. C. (1985). Disease control in the landscape. Co-operative Extention Service Ohio State University Bulletin 614. 25p.

Riffle, J. W., and G. W. Peterson. (Eds.). (1986). Diseases of trees in the Great Plains. USDA Forest Service Gen. Technical Report RM-129. 149p.

Shigo, A. L., and E. H. Larson. (1969). A photo guide to the pattern of discoloration and decay in living northern hardwood trees. USDA Forest Service Research Paper #NE-127. 100p.

Shigo, A. L. (1987). A new tree biology. Shigo and Trees Associates, Durham, NH. 595p.

Shurtleff, M. C., and R. Randall. (1975). How to control tree diseases and pests. Intertec. Pub., Kansas City, MO. 107p.

Sinclair, W. A., H. H. Lyon, and W. T. Johnson. (1987). Diseases of trees and shrubs. Cornell University Press, Ithaca, NY. 574p.

Smith, W. H. (1970). Tree pathology. Academic Press, New York. 309p.

Stipes, R. J., and R. J. Campana. (1981). Compendium of elm diseases. American Phytopathological Society Press, St. Paul, MN. 96p.

Streets, R. B. (1975). Diagnosis of plant diseases. University of Arizona Press, Tucson.

Tainter, F. H. (1979). Diseases of Arkansas forests. Arkansas Forestry Commission, 164p.

U.S. Department of Agriculture, Forest Service. (1979). A guide to common insects and diseases of forest trees in the northeastern United States. N.A.S.&P.F. publ. NA-FR-4. 127p.

USDA Forest Service. (1985). Insects and diseases of trees in the South. USDA General Report R8-GR5.

PERIODICALS

American Nurseryman
Arbor Age
Arboricultural Journal
Canadian Journal of Forest Research
European Journal of Forest Pathology
Grounds Maintenance
Journal of Arboriculture
Landscape Maintenance (formerly *Weeds, Trees and Turf*)
Phytopathology
Plant Disease (formerly *Plant Disease Reporter*)

Index